Women in Engineering and Science

Series Editor
Jill S. Tietjen
Greenwood Village, CO, USA

The Springer Women in Engineering and Science series highlights women's accomplishments in these critical fields. The foundational volume in the series provides a broad overview of women's multi-faceted contributions to engineering over the last century. Each subsequent volume is dedicated to illuminating women's research and achievements in key, targeted areas of contemporary engineering and science endeavors. The goal for the series is to raise awareness of the pivotal work women are undertaking in areas of keen importance to our global community.

More information about this series at http://www.springer.com/series/15424

Eucharia Oluchi Nwaichi
Editor

Science by Women

Stories From Careers in STEM

 Springer

Editor
Eucharia Oluchi Nwaichi
Biochemistry
University of Port Harcourt
Port Harcourt, Nigeria

ISSN 2509-6427   ISSN 2509-6435   (electronic)
Women in Engineering and Science
ISBN 978-3-030-83034-2   ISBN 978-3-030-83032-8   (eBook)
https://doi.org/10.1007/978-3-030-83032-8

This Springer imprint is published by the registered company Springer Nature Switzerland AG
The registered company address is: Gewerbestrasse 11, 6330 Cham, Switzerland

This book is dedicated to all male Champions of Gender in STEMM careers for the good work they do for a saner and justiciable space for women and men in the field of Science, Technology, Engineering, Mathematics, and Medicine (STEMM).

Foreword

It is a great honour to me that I was asked by Dr. Eucharia Oluchi Nwaichi, the President of Organization for Women in Science for the Developing World (OWSD) University of Port Harcourt (UNIPORT) Branch, to write a Foreword to a new book, *Science by Women*. The initiative to collate this book was inspired by amazing presentations made by noble women in various fields of expertise in science at monthly seminars held for the OWSD UNIPORT Branch, Nigeria Chapter, since inception of her (Eucharia) administration. To introduce myself, my name is Gertie Arts and I am a senior research scientist in aquatic ecology and ecotoxicology at Wageningen University and Research, The Netherlands. Since 2010 I am participating in SETAC (Society of Environmental Toxicology and Chemistry) governance, first as member of the SETAC Europe Council, later on as President of SETAC Europe, and member and President of SETAC World Council. Currently, I am past president of the World Council of this society. From my roles in SETAC I was very privileged to meet and get to know several authors of this book, especially the book chapter I co-authored, during SETAC Africa biannual and global meetings. In the scientific arena covered by STEMM, many professional societies are active at the continental and global level, SETAC being one of them. Through SETAC, I came into contact with STEMM.

There is no doubt that the potential contributions of women to science are immense and valuable. Female students often have an outstanding performance in their studies and PhD research, but after their graduation it is not always obvious they can follow the same scientific careers as men. Especially when taking care of a family, women always feel responsible and wish to find the balance among the responsibilities they have for the science and research, for their family and for their own health and personal needs. Therefore, a scientific career requires a constant balancing act of women. This book will give you an insight to how women in science perform, which challenges they encounter, how they overcome these and how they reach their goals in science and fulfil their submissions to servicing science careers for the sustainable development of the society.

I am immensely impressed by the valuable contributions in this book and I sincerely hope this book will inspire you in your personal scientific careers!

Past President, Society of Environmental Toxicology and Chemistry (SETAC)

Wageningen, The Netherlands   Gertie Arts,
2021

Acknowledgements

I have to start by thanking my amazing fan and husband, Chucks. From serving warm reminders to setting up virtual meetings where most presentations were made before conception as chapters, he was as important to this book getting done as I was. A million thanks, dear.

Thanks to everyone on the Books Series Committee of Organization for Women in Science University of Port Harcourt (OWSD UNIPORT) who helped to harness ideas into reality. Prof. A. I. Hart led this great team comprising Dr. B. Odogwu (Secretary), Dr. E. Fomsi, Dr. C. C. Ikewuchi, Dr. M. T. Nwakuya and Mrs. Martha Musa.

Prof. Nimi D. Briggs is greatly appreciated for reviewing some chapters within a short timeframe amidst his crowded schedules. Prof. Helen Imafido, Dr. E. O. Nwaichi, Dr. Ifeoma Chidebe, Dr. Ronald Kakeeto, Dr. Sally Chikuta, Dr. Eric Agoyi, Dr. Nwakuya and Dr. C. C. Ikewuchi are highly appreciated for painstakingly reviewing assigned chapters of this book.

Writing a book is more difficult than I thought and more rewarding than I could have ever imagined. The ever-committed members of OWSD UNIPORT were never-weary of saying 'you can do it' – I cherish your trust.

I'm eternally grateful to my mom, who loves reading everything written by me about me – Mom you spur me on! Thanks heartily.

My siblings always understand when I do not pick up their calls because I am paying attention to the manuscript – I appreciate you De Vin, Raymond, Adanne, Obinna, Anayo, Nneka, Ikechukwu and Ikemba.

Professor Joseph A. Ajienka, Prof. Stephen A.Okodudu, Prof. Mike Faborode, Mr. Vincent Onobun, Prof. Owunari Georgewill, Prof. Emeritus Emmanuel Okogbue Anosike (Late), Prof. Edward Obiozor Ayalogu (Late), Prof. L. C. Amajor (Late), Dr. Ejikeme Ugwoha, Prof. Onyewuchi Akaranta, Prof. Ndowa E.S. Lale, Dr. Justice Obinna Osuoha and Prof. V. C. Ukaegbu are highly acknowledged for inspiring me through various forms of support, encouraging words and availability to talk when I call.

How can I forget Ebenezer Oloyede, for all the technical support with my Laptop and accessories? Thank you Eben!

A very special thanks to Springer Publishers for agreeing to go on this journey with me and for their warm corresponding team members and useful suggestions all the way.

Contents

Contributors

Juliet Chinedu Alex-Nmecha Department of Library and Information Science, Faculty of Education, University of Port Harcourt, Port Harcourt, Nigeria

Gertie Arts Wageningen University and Research, Wageningen, The Netherlands

Chioma Blaise Chikere Department of Microbiology, University of Port Harcourt, Port Harcourt, Nigeria

Department of Environmental Sciences, College of Agriculture and Environmental Sciences (CAES), Florida Science Campus, University of South Africa (UNISA), Johannesburg, South Africa

Clara Chinweoke Ifeanyi-Obi Department of Agricultural Economics and Extension, University of Port Harcourt, Port Harcourt, Nigeria

Florence Onyemachi Nduka University of Port Harcourt, Port Harcourt, Nigeria

Eucharia Oluchi Nwaichi Department of Biochemistry, University of Port Harcourt, Port Harcourt, Rivers State, Nigeria

Exchange and Linkage Programmes Unit, University of Port Harcourt, Port Harcourt, Rivers State, Nigeria

Gloria Ukalina Obuzor Department of Pure & Applied Chemistry, University of Port Harcourt, Port Harcourt, Rivers State, Nigeria

Blessing Adanta Odogwu Department of Plant Science and Biotechnology, University of Port Harcourt, Port Harcourt, Nigeria

Ngozi Nma Odu Pamo University of Medical Sciences, Oyigbo, Rivers State, Nigeria

Linda Aurelia Ofori Kwame Nkrumah University of Science and Technology, Kumasi, Ghana

Linda Uchenna Oghenekaro Department of Computer Science, Faculty of Science, University of Port Harcourt, Port Harcourt, Rivers State, Nigeria

John Gibson Ogonu Department of Library and Information Science, Faculty of Education, University of Port Harcourt, Port Harcourt, Nigeria

Rosemary Nkemdilim Ogu Department of Obstetrics and Gynaecology, University of Port Harcourt, Port Harcourt, Rivers State, Nigeria

Temitope Olabisi Onuminya TETFund Centre of Excellence on Biodiversity Conservation and Ecosystem Management, University of Lagos, Lagos, Nigeria

Department of Botany, Faculty of Science, University of Lagos, Lagos, Nigeria

Uchechi Bliss Onyedikachi Department of Biochemistry, College of Natural Sciences, Michael Okpara University of AgricultureUmudike, Umuahia, Abia State, Nigeria

Blessing Minaopunye Onyegeme-Okerenta Department of Biochemistry, University of Port Harcourt Choba, Port Harcourt, Rivers State, Nigeria

Beatrice Olutoyin Opeolu Cape Peninsula University of Technology, Western Cape, South Africa

Rachel Paterson BetterManager, Professional Training & Coaching, San Francisco, CA, USA

Temitope Olawunmi Sogbanmu Ecotoxicology and Conservation Unit, Department of Zoology, Faculty of Science, University of Lagos, Lagos, Nigeria

TETFund Centre of Excellence on Biodiversity Conservation and Ecosystem Management, University of Lagos, Lagos, Nigeria

Memory Tekere Department of Environmental Sciences, Florida Science Campus, University of South Africa, Pretoria, South Africa

Eliane Ubalijoro Future Earth, Montreal, QC, Canada

Department of Geography, University of Concordia, Montréal, QC, Canada

Institute for the Study of International Development, McGill University, Montréal, QC, Canada

Ejikeme Obed Ugwoha Department of Environmental Engineering, University of Port Harcourt, Port Harcourt, Rivers State, Nigeria

Ijeoma Favour Vincent-Akpu Hydrobiology and Fisheries Unit, Department of Animal and Environmental Biology, Faculty of Science, University of Port Harcourt, Port Harcourt, Nigeria

About the Author

 **Juliet Chinedu Alex-Nmecha** is a librarian and a lover of books as she calls herself 'Book Dragon'. Her love for books started when she was employed as a Librarian-in-Training in March 2006 in the Donald Ekong Library of the University of Port Harcourt, Nigeria. She originally studied N.C.E. Social Studies/ Economics and BEd Economics. When she changed career prospects, she quickly bagged Master's in Library Science (MLS). Her quest for more knowledge spurred her into Master's in Education (Sociology of Education Option) to find out man's relationship to education as regards their immediate society and later PhD in Library and Information Science.Dr. Juliet C. Alex-Nmecha is a lecturer in the Department of Library and Information Science at the University of Port Harcourt, Rivers State, Nigeria. She has published articles in local and international journals, attended and presented papers in Library and Information Science conferences and other conferences related to Education. Dr. Alex-Nmecha does research on preservation and conservation of library resources, artificial intelligence in libraries, reference services, library advocacy, academic librarianship, and knowledge management.Dr. Alex-Nmecha is a member of some professional associations such as Nigerian Library Association (NLA), Organization for Women in Science for the Developing World (OWSD) University of Port Harcourt branch, Librarian Registration Council of Nigeria (LRCN), Association of Women Librarians in Nigeria (AWLIN) where she currently serves as the Financial Secretary and in Library Advocacy Group (LAG) too. She is an award winner on the Best Practicing Librarian in Rivers State, Nigeria, and Library Services Promotion award winner in Nigerian Library Association, Nigeria. She also had honours on Mentorship, mentoring younger colleagues in same profession. She is currently the chairperson, Nigerian Library Association, Rivers State Chapter. She lives in Nigeria.

Gertie H. P. Arts studied Biology at Radboud University in Nijmegen. She is Doctor in the Natural Sciences and received her title in Aquatic Ecology at the same University. She works at Wageningen Environmental Research as a senior research scientist in the Environmental Risk Assessment team. She is involved in higher tier experiments for higher-tier aquatic risk assessment procedures for contaminants (e.g. for the registration of pesticides). She works on the ecological evaluation of pesticide risks and on ecological and other aspects of freshwaters in the Dutch agricultural landscape, for example multistress and eutrophication. She has a main focus on aquatic macrophyte and terrestrial plant risk assessment, including research in the laboratory, mesocosms and field and developing and evaluation of new higher-tier methods. Her research focuses on the effects of contaminants on plants in different experimental settings and at different levels of biological organisation. Within the Dutch Pesticides research programme, she participates as a project representative of the theme 'Ecological Risk Assessment of Pesticides', dealing with the effects and ecological risks in surface water. In 2008 she was deputy project leader of the Nature Balance. For this project she worked three days per week at the Netherlands Environmental Assessment Agency in Bilthoven, the Netherlands. Because of her broad ecological and abiotic knowledge of aquatic ecosystems, she has been involved in the Natura 2000 reporting of aquatic habitats in the Netherlands and in projects within the context of the Water Framework Directive and its application. She participated in the Red List Project for the European Commission. Within this project she was member of the freshwater group and cooperated in this group to perform a red list assessment for freshwater habitats in Europe. She also participates in freshwater projects where Natura 2000 and the Water Framework Directive both play a role.She has been member of SETAC 2006 Scientific Committee. She has been co-chair of the Organising Committee of the AMRAP workshop focusing on Aquatic Macrophyte Risk assessment for plant protection products, and from 2009 until 2015 she was chair of the Steering Committee of the SETAC Advisory Group AMEG (Aquatic Macrophytes Ecotoxicology Group, since 2014 Plants Advisory Group). From 2010 to 2016 she was member of the SETAC Europe Council. From May 2014 until May 2015 she was vice president of SETAC Europe. From May 2015 to May 2016 she was president of SETAC Europe. From May 2016 to May 2017 she was immediate past president of SETAC Europe. In 2014 and 2015 she was co-chair of the Steering Committee of the non-target terrestrial plant workshops organized under the umbrella of SETAC. From May 2011 until January 2016 she was Plant Editor of *Environmental Toxicology and Chemistry*. From January 2018 to December 2018 she was vice president of SETAC World Council. From January 2019 to December 2019 she was president of SETAC World Council. Since 2020 she is Past President of SETAC World Council.She is the proud mother of two boys, now being scientists themselves. She loves running, walking in nature, working in her vegetable garden and travelling.

Chioma Blaise Chikere is a senior lecturer in the Department of Microbiology, University of Port Harcourt, Nigeria. Her specialization is Molecular Environmental/Petroleum Microbiology. She received Third World Organization for Women in Science (TWOWS), now OWSD, Postgraduate Fellowship in 2005 for split-site PhD programme at universities of Port Harcourt, Nigeria and Pretoria, South Africa, respectively. She has published widely with over 55 journal articles and more than 50 conference papers. Dr. Chikere was awarded International Foundation of Science (IFS) research grants in 2007 and 2012. Her PhD Thesis titled 'Bacterial Diversity and Community Dynamics During the Bioremediation of Crude Oil-Polluted Soil' was selected in 2013 by National Universities Commission (NUC) as the Best Doctoral Thesis in Biological Sciences for 2010. Other awards she received are: University of Port Harcourt 2015 distinguished merit award for diligent and meritorious service; TWAS-UNESCO Associateship (June 2016–December 2020) to the Department of Environmental Sciences, University of South Africa under Professor Memory Tekere; and the Elsevier Foundation 2017 Green and Sustainable Chemistry Challenge second prize held in Berlin, Germany. The Elsevier project has benefitted a Niger Delta Community that suffered untold environmental degradation occasioned by crude oil pollution. The polluted land in this locality has been fully recovered through bioremediation with robust ecosystem services fully restored. Dr. Chikere is an astute researcher and mentor having provided immense support and training to graduate students in her research group by providing them international exposure which they did not have previously. The International Foundation for Science (IFS) research grant she was awarded in 2012 sponsored two master's project students' researches from 2012 to 2015 in her department. She equally secured six PhD fellowships for her graduate students (four women and two men), and she supervised MSc programmes in Environmental Microbiology and Bioremediation in the Department of Microbiology, Uniport at the World Bank Africa Centre of Excellence in Oilfield Chemicals Research (ACE-CEFOR) University of Port Harcourt in 2014–2017, 2017–2020 and 2018–2012. Five of these students (three women and two men) have defended their PhD theses, and all have good records of journal publications and international conference papers (with travel grants from Dr. Chikere's membership in Society for Applied Microbiology – SfAM UK) from their researches, while one (Ms. Chidinma Okafor 2018 to 2021) is about to complete her PhD programme. Dr. Chikere has active collaborations in College of Agriculture and Environmental Sciences (CAES), University of South Africa, Florida Science Campus, South Africa for students' research visits with Professors Memory Tekere and Khayalethu Ntushelo (he has co-supervised two PhD students with Dr. Chikere and hosted one Emmanuel Fenibo for 2 months in his lab in 2019 while the second, Chidinma Okafor, is about to complete her research).She equally has competencies in public engagement and science communication to non-peers in science diplomacy, advocacy and mentorship of young career scientists especially for the retention of girls and women in STEMM subjects.

Chinwoke Clara Ifeanyi-Obi is a lecturer in the Department of Agricultural Economics and Extension, University of Port Harcourt, Rivers State, Nigeria. Currently, she is Research Fellow of the African Institute for Mathematical Sciences Next Einstein Initiative (AIMS NEI) Fellowship Program for Women in Climate Change Science. She is implementing a project titled Gender-Responsive Climate Change Adaptation Initiative in Nigerian Agriculture **(GCAINA)**.Clara holds an MSc in Agricultural Extension and Rural Sociology and PhD in Agricultural Extension from the Federal University of Technology, Owerri, Imo State, Nigeria, with specialty area in rural and community development. Her passion to help rural dwellers improve their standard of living is the motion behind studying agricultural extension.Her research interest is mainly on building farmers resilience to climate change impacts and strengthening their capacity for increased uptake of Climate Smart Agriculture (CSA). Her attraction to this area of research is the devastating effects climate change exerts on rural agriculture and her desire to help the rural poor farmers overcome this challenge. Dr. Ifeanyi-Obi is an emerging leading researcher in Climate Smart Agriculture (CSA) and has made significant contribution in providing evidence for policy decisions in agriculture and environmental issues in Nigeria. She has contributed immensely in building the capacity of other researchers through the numerous public engagement activities she has been involved in. She has published over 70 journal articles; presented research articles in over 20 conferences; and facilitated in over 28 training workshops including the Association of Commonwealth Universities Developing the Next Generation of researcher's workshops for Early Career Researchers in Nigeria.Dr. Ifeanyi-Obi was a post doc research fellow of the United Kingdom Department for International Development under the Climate Impact Research Capacity and Leadership Enhancement (Circle) Programme; a global fellow of Center for Human Rights and Humanitarian Studies, Brown University, USA; and a fellow of West African Science leadership Programme (WASLP). After her post doc, she was awarded two research uptake grants to facilitate the mainstreaming of Climate Smart Agriculture technologies and innovations identified through her post doctorate research into policy and practice in Nigeria. This confirms the high relevance of her research findings. She held a policy discourse and training working for farmers in climate change to the effect in 2017 and 2018 respectively. She is a member of eight professional associations within and outside her country, Nigeria. She is the National Coordinator of CIRCLE Fellows in Nigeria; Coordinator of Foresight and Climate Smart Agriculture Working group of the Nigerian Forum for Agricultural Advisory Services (NIFAAS); Rivers State Coordinator of Rural Sociological

Association of Nigeria; and Assistant Secretary of Organization for Women in Science for Developing countries, UNIPORT Chapter. She is an agribusiness consultant to two agri-organizations based in the United States, Alpha Zomax Consultants INC and Organisation for Development of Agriculture in West African Countries (ODAWAC). In recognition of her dedication and scholarly activities in teaching and research, she has won over eight scholarships and awards.Clara believes no research has taken place unless there is impact, hence her dedication to public engagement activities. She is the Founder of Vulnerable Lives Enhancement Foundation (VEF), a foundation targeted towards supporting vulnerable groups particularly, women, youth and children to enhance their livelihoods. She has done various volunteer works particularly in the area of hosting free career day for government-owned schools. An avenue she uses to counsel and guide young girls into choosing STEM career. Clara enjoys family life and dedicates her spare time to family get-together with her husband and three charming children.

 **Florence Onyemachi Nduka** is Professor of Parasitology engaged in teaching at both the undergraduate and postgraduate levels and research in the Department of Animal And Environmental Biology in Faculty of Science, University of Port Harcourt, Rivers State, Nigeria. Her research focuses on the epidemiology of major parasitic infections found in the tropics such as malaria, schistosomiasis, soil-transmitted helminthiasis and filariasis. Her research outputs have provided baseline data, validated diagnostics, shown community perception and practices that enhance risks, role of environmental degradation in transmission and advocacy and public health education in the reduction of infections. Her research has also investigated the compliance levels to WHO prescribed malaria preventive therapy in pregnant women in different sections of the country.She led the mapping of schistosomiasis and soil-transmitted helminths as a Consultant to the country's ministry of health in southern states of Nigeria in 2014. This project was anchored by Sightsavers Donor partner supported with funding by CIFF United Kingdom. The study results revealed that all the States and the Federal capital of the country were endemic for one or both diseases and that 359 of the 774 local government area needed interventions to ameliorate the effects of the diseases. The data generated from this essential survey provided much needed evidence for appropriate and sustainable intervention and policy formulation by the central government in the control of neglected tropical diseases (NTDs). Following the recommendations of this study, there has been a scale up uninterrupted provision and administration of recommended medicines used in preventive chemotherapy of these infections.Together with her team of earlier career researchers they demonstrated the role of abandoned quarry pits filled with waters in the transmission of schistosomiasis and the need to amend existing laws that can make it mandatory for miners to restore the environment after degradation. These pits are formed after the

blasting of quarry stones used in road constructions and building projects. When abandoned, they get filled with water and form recreation centres for the communities given their nearness to the homesteads when compared to natural river sources. In visiting the water sites for washing of clothes and farm produce, swimming and fishing, more individuals get infected with this water contact infection. Public enlightenment to the communities organized by her team led to information on the nature and source of infection, the need to stay away from the water source and report to health facilities when symptoms are noticed and a steady decline in the levels of infection. The local governments of the areas were encouraged to provide portable waters for the communities.Professor Nduka has demonstrated through sustained research the need for vulnerable malaria group, the pregnant women to adhere to the WHO preventive treatment of taking recommended doses of sulphadoxine-pyremethamine as direct observed treatment during ante-natal care in health facilities. Her research showed increased malaria parasites presence in the placenta of infected women than in the peripheral blood and lowest infection levels were observed in the group of pregnant women who adhered to and completed the therapy. She also showed low compliance levels were common occurrence due to varied reasons from healthcare workers' attitude to late commencement of ante-natal visits and urged the relevant officials to put measures in place to upgrade compliance. Her recent observation has shown marked improvement in compliance in Rivers State of Nigeria.She sits as a member of the National Steering Committee on Neglected Tropical Diseases and formulates policies for implementation of control measures especially the water, sanitation and hygiene components to the underserved communities at high risk of these infections. She is also the Chair of the Board of the Niger Delta Development Commission Research on Malaria and Phytomedicine, and her team has recently concluded evaluation of malaria infection in pregnant women and children 0–5 years in three states of the Niger Delta. She is currently involved in understanding the reasons for the low COVID-19 impact in her section of the country and wondering if it has anything to do with malaria endemicity; work is ongoing on sero-prevalence of COVID-19 infections and malaria prevalence.Professor Nduka's fascination with science was cut at the feet of her late father, a dentist who introduced her to the world of mathematics and the sciences very early in life. She watched him set and mould dentures and was touched by the 'miracle' of the minted denture and the joyful smile of the patient. Through her father she learned to read diverse books from literature to physics, and one of her favourites was 'The Giants of Science' capturing the magic of discoveries by world greats. She was finally sold on research in her third year in the university (1979) when she was taught by two women fresh PhDs in Parasitology who set the tone to her future accomplishments. She has followed in their footsteps and mentored many others, men and women some of whom are professors holding their own grounds in research. She is a strong advocate for girls in STEMM and was recently interviewed by the AGE group for Girls in STEMM. She has presented lead papers on gender gaps and violence against women. She has held different administrative and leadership positions in the university system and her community. She is married to a very supportive husband, a professor of Statistics, and they have four children with two daughters pursuing PhDs in STEMM.

Eucharia Oluchi Nwaichi is a faculty member in the Department of Biochemistry and the Acting Director of Exchange and Linkage Programmes Unit, both of the University of Port Harcourt, Nigeria. Her research focuses on monitoring and understanding the quality of soil, water and air environments and looking to using her patent and associated technologies to remedy impacted environment. Her primary areas of interest include phytoremediation of impacted soils and water, including wetlands, remediation by natural attenuation and air quality monitoring, while her other research interests include toxicological assessments of exposed foods and human population. Her current projects range from production and characterization of enzymes for enhanced treatment of organic pollutants in petroleum refinery effluent to attenuation of petroleum hydrocarbon fractions using rhizobacterial isolates as well as optimization of bioremediation-cocktail for soil clean-ups. She spent several years in industry working on quality control and assurance systems and facilitated HSE compliance. In the past, she has worked with the food and beverage as well as the oil and gas industries. She completed her PhD in Biochemistry at the University of Port Harcourt and is affiliated with the Next Einstein Forum, West African Science Leadership Programme, International Society of Engineering Science and Technology, Academic Society for Functional Foods and Bioactive Compounds, American Chemical Society, Rivers state Child Protection Network, International Phytotechnology Society, Nigerian Society of Biochemistry and Molecular Biology, Organization for Women in Science for the Developing World, Caritas Empowerment and Development Initiative, International Society for Environmental Technology and Nigerian Institute of Management.Eucharia is the sixth of nine children who all pursued graduate degrees in various disciplines. Her parents celebrated recorded successes among the children, whenever they came – for example a new pair of shoes, a dress and a hat always went to Eucharia at the end of third terms (when results are declared) when she clinched First position in her primary school days. Again, she was encouraged to teach Maths and Chemistry to her older siblings and classmates in the rural schools during the breaks. That is the kind of deep support that emboldened and propelled her to the top of her field. Inspired early to use her imagination and initiatives to solve problems, she was captivated by any project that involved taking initiatives and helping communities. Today, the use of less harmful technologies to cleaning contaminated environment has become part of her mission to create an informed and participating community in environmental management. Eucharia has a patent to her credit in phytoremediation. As an Environmental Biochemist, she leads a research team of policymakers and academic collaborators who synthesize evidence to take on dearth of evidence-informed policymaking.Among other notable achievements, she is the pioneer President of Organization for Women in Science for the Developing World (OWSD) University of Port Harcourt Branch where she has grown the membership to over one hundred and seventy (170) since 2019.

Eucharia has just been elected as the National Vice President of OWSD Nigerian National Chapter. She promotes Evidence-Informed Decision-Making, EIDM. For example, given her knowledge and expertise in Chemical Safety and Security, she created an enabling environment for researchers, policymakers and industry partners to be trained on Chemical Safety and Security while using local evidences to push policy and researches around production, use, importation, exportation and storage of chemicals, to ensure they are not used as weapons. It is important to state that she has successfully self-funded and rolled out four annual editions of such laudable gestures (Seminar Series) since year 2017. A government agency following the second edition invited her to their laboratory to do chemical risk assessment and output was a redesign of the Chemical Store. She promotes green and sustainable chemistry and is a great science communicator who is poised with influencing policies with her science and increasing the space for women and girls in science. Her patent, findable recognitions, awards and honours, including 2020 Evidence Leader first Runner-up, 2019 Fellow of Next Einstein Forum, 2016 Affiliate of the African Academy of Science, 2015 Fellow of Commonwealth, 2015 University of Port Harcourt Merit Awardee for numerous contributions, 2015 Honours roll of the Faculty of Science of the University of Port Harcourt and 2013 Fellow of UNESCO L'Oreal for Women in Science, among others, make her a great influence on women and girls in science and she has successfully mentored many. She features on radio and television stations to discuss issues around, girls, women and science. She has published her works in > 75 reputable journals and has well-rated book chapters. She has post-doctoral experiences in the USA, the UK and Poland and has presented papers and served in various committees and panels, locally, regionally and internationally. Her husband is her biggest fan and rallies round to see keeps to career targets.Leadership magazine in year 2017 described Eucharia as One of Nigeria's shining lights in the Sciences. Her work has been praised by Silverbird Television as they recognized Eucharia as one of '16 Prominent Nigerian Women That Excel in Science and Research'. These recognitions are findable on Google. In her spare time, she is an actress and has starred in three movies and six skits. She enjoys swimming, dancing and meeting new people.

Gloria Ukalina Obuzor , Professor of Chemistry, is the first female President of the Chemical Society of Nigeria in its 37 years of existence in 2014. She obtained her BSc in Chemistry from the University of the District of Columbia (UDC), Washington DC, USA, in 1979 with a Rivers State scholarship; an MSc from the University of Port Harcourt (UPH), Port Harcourt, Nigeria, in 1989 with a UPH scholarship; and a PhD in Organic Chemistry with a focus in Organometallic Chemistry from the University of Manchester Institute of Science and Technology, Manchester (now called the University of Manchester), UK, in 1998 with a Commonwealth Scholarship. Prof. Obuzor is the Vice

President, Federation of African Societies of Chemistry (FASC); is a member of Chemical Society of Nigeria, American Chemical Society (ACS), Royal Society of Chemistry, Bureau of International Union of Pure and Applied Chemistry (IUPAC), Committee of Chemistry Education of IUPAC and the Inorganic National Adhering Officer of IUPAC; was an associate member of International Activity Committee of ACS; and was a member of Division of Chemistry and Environment of IUPAC. Her research interest focuses on natural products, search for unusual phytochemicals among the rich flora of the Niger Delta, Nigeria, and the production of fruit wine from locally available fruits. As a chemist, she strongly upholds the tenet that every chemical equation is a potential industry, and she has applied the equation $C_6H_{12}O_6 \rightarrow 2C_2H_5OH + 2CO_2$ for the production of Ukalina Fruit Wines. Gloria Ukalina Obuzor who has several publications with three Nigerian patents is a role model and a mentor to many.

Blessing Adanta Odogwu is a lecturer in the Department of Plant Science and Biotechnology in the Faculty of Science, University of Port Harcourt, Nigeria. She is a plant breeder currently working on improving the culinary qualities and adaptation capability traits of indigenous legumes in Nigeria. During her research, she will be investigating the genes responsible for these traits using genomics and bioinformatics tools. She will also be interacting with various stakeholders of the legumes value-chains to document their perceptions and include them in the participatory selection of the legume varieties with their choice traits. She is a member of the Organization for Women in Science for the Developing World (OWSD) and an ardent advocate of women and gender mainstreaming in research and technology uptake. She is a voluntary mentor of youths interested in agriculture science on the Grooming Leaders in Agriculture (GLA) platform.Growing up, Dr. Odogwu remembers from childhood how her mother grew vegetables between rows of corn and cassava in their garden and would caution her and her five siblings not to disturb the plants and their seeds. But only as an adult, while studying plant sciences and genetics, did she fully appreciate the importance of seeds, including the role improved seed can play in bettering the lives of smallholder farmers. However, Dr. Odogwu's career journey in plant breeding started when she got admitted into the Department of Plant Science and Biotechnology at the University of Port Harcourt to study basic plant sciences. At the completion of her study, she was employed as a graduate assistant in the Department. She was fascinated with genetics and plant taxonomy courses, so at the master's level, she studied plant biosystematics and taxonomy. She became intrigued with the application of biotechnological tools in the manipulation of genes for plant improvement and so she decided to study plant breeding for her PhD. She was awarded a full scholarship by the Regional Universities Forum for Capacity in Agriculture (RUFORUM)/Carnegie Cooperation of New York to study Plant

Breeding and Biotechnology at Makerere University in Uganda.During her doctoral research, she saw the need to enhance her transferrable skills in plant breeding that will enable her develop technology that will improve the lives of farmers. In 2014, she applied for a career development fellowship offered by the African Women in Agricultural Research and Development (AWARD). The fellowship helped her to properly develop a career path to enable her reach her long-term goal of attaining to institutional positions of influence and having an effective breeding program. Her short-term goal was not only to complete her doctorate degree but to get the requisite and transferrable skills in plant breeding, genomics and bioinformatics. Her research was conducted at the legume breeding program at the National Crop Resources Research Institute (NaCRRI) and CIAT, Uganda, where she gained experience on how to develop a legume breeding program and enhance her skills in the use of molecular markers in genomic selection. At the same time she won the Norman Borlaug Leadership Enhancement in Agricultural Program fellowship that supported her to visit the labs of Dr. J. Kelly, a bean breeder at Michigan State University, USA, and a plant pathologist, Dr. J. Steadman of the University of Nebraska, USA. While visiting their labs, she saw the need to enhance her genomic and bioinformatics skills using high-end technology, so she applied for the African Biosciences Challenge Fund fellowship which was carried out at the BecA-ILRI Hub, Nairobi. These fellowships helped her to complete her doctoral study and develop excellent networks.On her return to her home institution, Dr. Odogwu started working on her goal of establishing a legume program and mentoring youths in agriculture through the Grooming Leaders in Agriculture (GLA) platform. She also got funding from a post-doctoral fellowship by RUFORUM/and Carnegie Cooperation of New York and IFS in 2018. The funding enabled her to start a legume breeding program and integrate into the university system and supported the research work of six postgraduate students which included two international students.Dr. Odogwu is married with two children. When she's not at work, she enjoys reading novels, traveling and volunteers her time in the public preaching work.

Ngozi Nma Odu is Professor of Food and Public Health Microbiology at the PAMO University of Medical Sciences, Port Harcourt. Professor Odu had also served the University of Port Harcourt both as a lecturer in the Department of Microbiology and as the first Acting Executive Director of the University of Port Harcourt Foundation. She is also a visiting professor to the Rivers State University. She obtained her first honours degree in Microbiology from Paisley College of Technology, Renfrewshire, Scotland (1979). While in Scotland, she received the Best Academic Prize for 2 consecutive years (1977 and 1978 respectively). She also holds a Master's degree in Food Science and Management from the prestigious University of London (Queen Elizabeth College, 1983). In 1989 she obtained a PhD in Microbiology from the University of Port Harcourt that made her the first PhD graduate in that department. She also attended a Leadership course at the prestigious John Hopkins University Baltimore, USA, 2005.She has had a varied working experience. She rose through the ranks in the Rivers State Ministry of Health to the post of Permanent Secretary, a post she held for six and half years (1999–2006) making her the longest serving Permanent Secretary in the Ministry. During her tenure as Permanent Secretary, the Ministry received accolades from WHO, UNICEF and Federal Ministry of Health (FMOH) for best practices in the area of immunization and Baby Friendly Initiative Programmes. Prof. Odu led the rescue team of the December 2010 ill-fated Sosoliso air crash (there were only two survivors). In January 2006, she was elevated to a position of a Commissioner in the Rivers State Civil Service in the Ministry of Education. As a commissioner, she made some remarkable contributions to Human Capital Development. She facilitated the award of overseas undergraduate scholarships to over 400 Rivers State students in Russia and Malaysia, and they have all graduated including 38 medical doctors. In the same vein, she also facilitated the award of overseas postgraduate scholarships to 75 Rivers State men and women in the USA, the UK and Europe, and she also initiated school beautification programme in public schools in the state.She had served her country in various ways. She reviewed three chapters of the 2004 (2nd Edition) of the National Health Policy of Nigeria. She co-reviewed the 2004–2007 Health Sector Reform Programme of the FMOH. She also appraised the Public Private Partnership (PPP) implementation on the health sector in Akwa Ibom State in 2005. She produced the draft State Strategic Health Development Plan (SSHDP) for Bayelsa State in 2009 as a consultant to the FMOH. She has served as a consultant to SPDC, TOTAL, NLNG and UNDP. She has partnered with global organizations such as Pathfinder International and Ford Foundation. She also served as State President of NCWS – National Council of Women's Societies (2001–2006). Under the NCWS umbrella, Prof. Odu led partnership with other professional and non-professional women organizations, government and philanthropists engaged in the following youth and women empowerment programmes: Economic empowerment of rural women (from 2003 to 2004, of which 2300 women benefited from several skills acquisition trainings); safe motherhood campaign for women in rural communities that trained many traditional birth attendants (TBAs – Medical Women Association (MWA) and nurses in Rivers State were facilitators during this exercise – 2000 women benefitted from this project); Federation of female lawyers (FIDA)-led free legal services for vulnerable women to abate gender-based violence. The NCWS also partnered with the Adolescent Project initiated by her Excellency, Hon. Justice Mary Odili JSC former First Lady and the wife of Dr. Peter Odili, former Governor of River State for youth empowerment in the State.Professor Odu was invested as a Fellow of NIM in July 2011. She was co-opted into council in 2014. She has served the institute in various capacities which includes among others Chairman of Board of Examiners, three times member of BOE, member of programmes committee, Chairman R/S NIM fund raising committee for the rehabilitation of the Management house.She has mentored/supervised many postgraduate and undergraduate students. She has published over 88 peer-reviewed articles in renowned international and national journals. She is an International Scholar and an award winner of the Keystone Symposia on Molecular and Cellular Biology 'X8-2013: Tuberculosis: Understanding the Enemy's'.Prof. Odu is a member and fellow of the Association of Quality Education in Nigeria. She is a member of Nigeria Society for Microbiology (NSM), Nigeria Environmental Society (NES), Nigeria Institute of Food Science and Technology (NIFEST) and American Society for Microbiology among others. She is a Peer Reviewer of the *Journal of Microbiology Research, Scientific & Academic Publishing, Journal of Public Health Research, Food and Public Health* and *Frontiers in Science.* She has attended several conferences and workshops, nationally and internationally, and is a resource person to the Federal Ministry of Health.She is a widow of late Dr. Chimdi A. Odu, and is blessed with 5 grown children and 12 grandchildren.

Linda Aurelia Ofori is a senior lecturer in the Department of Theoretical and Applied Biology, Kwame Nkrumah University of Science and Technology, Kumasi, Ghana.She is a microbiologist with a research interest in molecular bacteriology and infection. She has spent over ten years teaching various courses in microbiology at both graduate and undergraduate levels and researched various microorganisms affecting humans, animals, food and the environment.She has a BSc in Biological Sciences and an MSc in Environmental Science, both from KNUST, Kumasi, Ghana, and a PhD in Molecular Bacteriology and Infection from the University of Copenhagen, Denmark.Linda is the oldest of six daughters who pursued graduate degrees in science, engineering, and mathematics. Her parents raised their daughters to believe they could do anything. That's the kind of deep support that propelled her to the top of her field. Inspired early to use her imagination and be health-conscious, she was captivated by any project that involved human life and living it. Today, the study of microorganisms is the centrepiece of her mission to uncover mechanisms behind some drug resistance.As a researcher, she has led or partnered with research teams to work on diverse projects, including an IFS project on the presence of protozoan parasites on vegetables. Challenge Program for Water and Food and the Food and Agriculture Organization project on wastewater irrigated vegetables in the Volta Basin. She also worked on a DANIDA-funded project on antimicrobial drug monitoring and evaluation of resistance which led to her PhD. She is currently co-leading work on the genetic adaptation of non-typhoidal *Salmonella* in children under five and animals within communities in Ghana and Tanzania. This work focuses more on identifying the organisms and their susceptibility to antimicrobials. If these organisms are resistant, the mechanisms of resistance are investigated to inform policy. Research has earned her several publications in reputable journals.Linda is a strong advocate for career development and has mentored several scientists. She is active in educational outreach, including teaching and mentoring teaching and research assistants. She is a proud member of the

Women in Science Technology Engineering and Mathematics (WiSTEM). She has been a mentor for young girls in both first- and second-cycle schools in Ghana as part of the WiSTEM agenda. We inspire these young ones to pursue programs in STEM and allow them to apply science in their everyday lives. She is a mentor for university students on two programs supported by the Millennium Developing Authority (MIDA) and Master Card Foundation.She has been part of the Ghana Science Association for several years and has supported the agenda of promoting everyday science for young people in our schools. She is a proud fellow of the Ghana young academy.Serving as an academic advisor for students earning graduate and undergraduate degrees is part of her normal routine. She truly values the importance of mentorship in her success, giving credit to her parents as early influencers since they inspired imagination and a love of science. At various levels in school, her research and academic advisors were also inspiring and influential, further helping her to build self-confidence.Besides science, Linda is a warden of one of the traditional residence halls on the KNUST campus, which houses close to 800 students and serves on various committees.Today, Linda's husband, Martin, is her biggest supporter, and her four children, her greatest motivation.She likes to spend time with her children in her spare time, watch movies, cook or swim.

Linda Uchenna Oghenekaro is a lecturer and researcher in the Department of Computer Science at the University of Port Harcourt, Nigeria. Her current area of research involves mining business processes using various machine learning tools and techniques. Having started her career in the financial institution domain, she developed the knowledge of how business processes can influence the overall outcome of organizations, and the need for them to be discovered, monitored and enhanced. Her present career in academics has given her opportunity to research on various approaches that could be adopted to improve business processes.In 2014, she started her academic career as a graduate assistant in the Department of Computer Science. Working in the academia gave her the opportunity to explore various research areas within computer science domain. Her primary research area of interest include data science, text mining, machine learning, cloud computing, process mining and hierarchical temporal memory, while other research interest include cyber security and distributed database. Despite being a young faculty member with about six years of experience in tutoring and researching, Linda has positively impacted her colleagues, to always strive for excellence, and she has also mentored a lot of young computer scientists to make their presence felt, in this world of so many opportunities. She continues to improve herself academically, by attending workshops, presenting her papers at conferences, and producing journal articles among other academic activities.Linda grew up in Lagos with her parents and five sisters. Being only girls at the time, her parents supported and encouraged every one of them to have big dreams and achieve them. Despite having little finances, her parents trained all six girls to tertiary level of education, of which they have become professionals in their various fields of study. Linda was trained to understand that no house chore was made for any specific gender, but that everyone must contribute to making family life beautiful, and this approach to family translated to her approach to life in general, as no task should be left undone for the reason of gender. She obtained her BS in Computer Science from the University of Port Harcourt in 2009, and finished as the best graduated female student of the Department. In 2012, she joined the UBA Group where she managed their business processes and supported their online applications, both home and abroad. In 2014, she returned to academic world, and while working, she started her graduate program. Linda recently concluded her doctorate program in the Department of Computer Science. She is a member of the Organization of Women in Science for the Developing World (OWSD) and other professional bodies such as Nigeria Computer Society (NCS) and Computer Professional of Nigeria (CPN).She is married to a medical surgeon, who is also a researcher, and they are blessed with three children. At her leisure, she reads and sings with her children. She loves to cook traditional dishes, watch movies and have fun.

Rosemary   Ogu ,   Consultant   Obstetrician/ Gynaecologist, University of Port Harcourt Teaching Hospital (UPTH), Port Harcourt, Nigeria, is a Maternal Health Advocate with interest in Sexual/Reproductive Health and Gestational Diabetes Mellitus.In 1990, she won the National Association of Women in Science and Technology Prize for Best Science Girls in Nigeria. This spurred her to study Medicine and Surgery. She graduated (MBBS) with distinction from the University of Benin, Benin City, Nigeria, in 1999. Her residency training in Obstetrics and Gynaecology was from 2003 to 2008 at the University of Port Harcourt Teaching Hospital. She qualified with the fellowship of both the West African College of Surgeons and the National Postgraduate Medical College of Nigeria (FWACS, FMCOG).Rosemary Ogu has special research interest in the areas of sexual/reproductive health and non-communicable diseases. She is the Secretary of the World Health Organization/Federal Ministry of Health Maternal-Newborn-Child-Health Implementation Research Team for the reduction of maternal and perinatal mortality in Nigeria. She is the Principal Investigator of the UPTH World Health Organization Safe Childbirth Checklist Collaboration. As a dedicated Medical Woman and Past President of the Medical Women's Association of Nigeria, Rivers-State Branch, she collaborates with relevant government agencies and non-government organizations to advocate and offer sensitization, awareness, screening, preventive and care services for women's health.Born March 25, 1973, in Delta State, Nigeria, Rosemary Ogu became a lecturer 1 in October 2009 and Professor in October 2018. Between 2015 and 2018, She obtained a master's (MSc) in Reproductive Health from the World Bank supported Centre of Excellence in

Reproductive Health Innovation (CERHI), University of Benin, Nigeria. She has received numerous Scholarship and honorary awards including the 1990 National Association of Women in Science and Technology Prize for Best Science Girls in Nigeria; the 2009 Nigerian Medical Association (NMA) Commendation for Service to Rivers State; the 2014 Port Harcourt University Medical Students' Association (PUMSA) Merit Award as Clinical lecturer of the year, the 2015 Nigeria Medical Students Association Award for Active Involvement and Promotion of Reproductive Health in Women and Youths and the 2017 Medical Women's Association of Nigeria (MWAN) National Award as MWAN Jewel for outstanding Contribution to MWAN for achieving the United Nations Economic and Social Council (ECOSOC) Status.Prof. Ogu has attended numerous writing and communications workshop including Strategic Communications Workshop; Documenting Your Work – Telling Your Story at Ford Foundation Office, Lagos, Nigeria, organized in 2015 by Communications Consortium Media Centre, Washington DC, USA; Big Data for Better Science: Technologies for Measuring Behaviour Workshop held at the Carlton House Terrace, London, SW1Y 5AG, organized in February 2019 by the Royal Society.She has contributed to two WHO publications titled: *Care of Women and Girls Living with Female Genital Mutilation: A Clinical Handbook*, Geneva: World Health Organization (2018), Licence: CC BY-NC-SA 3.0 IGO; and WHO Safe Childbirth Checklist Implementation Guide: Improving the Quality of Facility-Based Delivery for Mothers and Newborns, World Health Organization (2015) (NLM classification: WQ 300), Geneva, Switzerland. ISBN 978 92 4 154945 5. She has numerous book chapters and over 60 published articles in peer-reviewed journals.Prof. Rosemary Ogu is married to Hillary Ogu and they currently reside in Port Harcourt. She continues to advocate for more women and girls in science and serves as a role model for upcoming scientists. She loves giving health talks, reading and travelling.

Uchechi Bliss Onyedikachi is an Environmental Biochemist, researcher and lecturer at Michael Okpara University of Agriculture, Umudike, Nigeria. Her professional goal seeks measures to access and ameliorate environmental pollution through environmental monitoring, assessment and implementation of various innovative methodologies to remediate hazardous effects of pollution to man and his environment. She has led her research team through couching, communication, motivation and collaborations to assess hazardous levels of heavy metals and carcinogenic hydrocarbons in various crude oil spilled sites like the popular Ogoni land in Nigeria, whose agricultural zones and water bodies have been devastated with oil spills as a result of oil exploration. Her interest in environment pollution assessment spurred her team into a bigger project done to assess toxicity levels of heavy metals and polycyclic aromatic hydrocarbons in foods grown in agricultural zones around industrial areas in the five states that make up the South East geopolitical zone of

Nigeria. She has assessed the health risk impact associated with consumption of various classes of foods cultivated in the contaminated areas of her studies. This research spanned further to her study on the cardio-toxic effect of heavy metals with particular interest on hexavalent chromium due to its prevalence in the foods assessed in selected agricultural areas located in industrial areas and most importantly the rising cases of cardiovascular diseases and sudden deaths reported in the south east zone of Nigeria. She is currently studying the proficiency of essential oils from plants/herbs grown in her locality in the bid to solving environmental toxicity health-related issues. This study would help in the development of guideline and health protection policy for industrial workers. Some of her works have been published in peer-reviewed journals with many more under review. She has also participated in the assessment of various waste dumps and also studied the phytoremediation effect of little ironweed bioaugmented with pig dung in soils impacted with spent oil; all these studies were aimed at developing innovative skills, technology and workable methods through which pollution can be mitigated thereby creating a cleaner, healthier and wealthier environment. She has presented some of her research findings in national and international societal conferences. This includes the 36th national conference (2017) of the Nigerian society of biochemistry and molecular biology conference held at the University of Uyo, Nigeria, where her research presentation won the award of the best presentation at the environmental toxicology plenary session on the topic health risk assessment through the consumption of food crops grown in the quarry site at Ishiagu, Ebonyi State, Nigeria. This was a breakthrough and a feat so desired by many of her colleagues.Uchechi is the first of six children born to the family of Raymond and Maria, Nwosu, who are disciplinarians known for their high moral standards. This core value nurtured her and her siblings into hardworking and focused scientists who believe in their abilities. She is also married and blessed with children. Her keenness to learn and share the knowledge she has amassed stirred her into launching an online platform on WhatsApp called 'YOU TOO CAN MAKE IT', where inspirational messages, jobs, scholarship opportunities, academic/career mentorship, experiences, successes, challenges and solutions are discussed among young people. Some of the participants in this forum have gained admission into higher institution via inspiration and mentorship, some have written books, and some have gained scholarships which gave them the opportunity to explore other parts of the world. Some others have turned useful by conquering peer pressure and overcoming their fears which caused them to lose focus in their visions especially their academics, etc.Her research dreams are to learn, build connections with associates and collaborate with other researchers, STEM leaders and professionals around the world thereby developing her knowledge, skills and capacity in statistical analytical proficiency, research innovation, communication and writing, and most importantly, to access the latest technological advancement in green technology that would assist in remediation and the establishment of waste management system and genetic engineering. This will assist to environmental remediation by improving existing waste management and renewable energy systems thereby creating job opportunities, improving health, agricultural yields and scaling business opportunities. She intends to build a world-class team which will include scholars and artisans in the future whose aim will be to recycle various waste products into reusable energy and other resources like bioremediation fertilizers. This could increase agricultural yield and as well assist in cleansing of our polluted environment.She enjoys spending time with her family, reading, and writing. She finds so much joy in fellowshipping and teaching believer's foundation knowledge at her local church where she teaches and serves as an ordained worker.

Memory Tekere is Full Professor at the University of South Africa (UNISA) and currently serving as Research Professor in the Department of Environmental Science. She researches extensively in Environmental Biotechnology and Microbiology focusing on microbial diversity and bioprospecting in different habitats particularly extreme and polluted environments, bioremediation of pollutants, water quality as well as solid waste management. Her current and recent projects include metagenomics and bioprospecting studies of acid mine drainage, landfills, carwash effluents, thermal springs and indigenous Afrotemperate forest biome. She has worked on funded research projects on microbial deep carbon life and fungi in treated drinking water and implications for public health.With a Biochemistry Honours degree and a PhD in Science (Environmental Microbiology) from the University of Zimbabwe in collaboration with Lund University, Sweden, she began her full-time academic career as a lecturer in Biological Science at the University of Zimbabwe in 2002 before moving to South Africa in 2008. She worked as a senior lecturer at the University of Venda in 2008, and in 2009 she moved to the University of South Africa, where she rose through the academic ranks from senior lecturer, associate professor to full professor by 2014. Teaching and areas of research cover aspects of environmental and industrial microbiology and biotechnology, bioremediation, water quality, ecotoxicology and waste management. Due to her notable and sustained performance in research, she is currently seconded to the position of a research professor.Memory was born in a small-scale rural farming community of Chenjiri, Mashonaland west province in Zimbabwe, and did her primary and high school education at disadvantaged rural schools and mission boarding schools respectively. She is a fifth born in a family of seven, four boys and three girls. Born to a father who was a teacher, good grades were expected always, and this propelled her to go all the way through the educational and professional ranks, seeing no limits or excuse for failure. Science resonated with her more than any other fields of study, and from an early age she felt passionate towards microbiology. As young girl in her environment, she tried to explain basic science around microbial food fermentations and spoilage, human and animal diseases, antibiotics, and unknowingly various aspects of microbial ecology and beneficial roles in nutrient recycling and pollution clean-up.In her academic career which spans 20 years, she has developed herself as a prolific academic excelling in her research, academic citizenship and mentorship. She is well published in journals, books and conference proceedings.

She is a rated researcher under the National Research Foundation, South Africa, being recognized as an established researcher in her field of research. In 2018 she was recognized by her institution, bestowing on her the competitive Woman in Research Leadership Award. She has enjoyed a fair share of research funding and training awards that include research funding from Water Research Commission, South Africa (2016–2019); Sloan Foundation (USA), 2015; National Research Foundation (SA) – THRIP grant, (2011–2013); UNISA Women In Research Grant, 2013–2015; Swedish Agency for Research Corporation with Developing Countries, (SAREC) – PhD funding; the United Nations Centre for Technology Advancement in Developing countries (UNCTAD); and Organization for Social Science Research in Eastern and Southern Africa (*OSSREA*), among others. She has provided mentorship to many including emerging researchers in her field, including to fellow junior academics, visiting research fellows, postdoctoral fellows, and graduating in excess of 29 MSc and PhD postgraduates by research to date.Memory is also a registered professional natural scientist with the South African Council for Natural Scientific Professions. She participates as a member in professional bodies such as American Society for Microbiology (USA), Society for Applied Microbiology (UK), South African Society for Biochemistry and Molecular Biology, Water Institute of South Africa and South Africa Bureau of Standards/ISO water quality subcommittee. She does regular reviews for several articles submitted to scientific journals: *Journal of Hazardous Material, Applied Microbiology and Biotechnology, Frontiers, Bioresources, PLOS ONE, Biomass and Bioenergy, British Journal of Applied Science & Technology, Journal of Pure and Applied Microbiology, Environmental Advances, Environmental Science and Pollution Research, Ecological Indicators* and *Process Biochemistry*, among others.Memory is a devout Christian, she likes spending her free time listening to gospel music, sermons and reading related books. Nature soothes her and she spends her time relaxing and taking walks in local nature reserves. She is a mother to an 18-year-old college-going son, who adores her and appreciates her hard work in science and as a mother.

 **Éliane Ubalijoro** , PhD, is the Executive Director of Sustainability in the Digital Age and the Future Earth Montreal Hub. She is Professor of Practice for Public-Private Sector Partnerships at McGill University's Institute for the Study of International Development, where her research interests focus on innovation, gender and sustainable development for prosperity creation, and her teaching over the last decade has focused on facilitating leadership development. She is also Research Professor at Concordia University in the Geography Department. She is a member of Rwanda's National Science and Technology Council. Eliane has been a member of the Presidential Advisory Council for Rwandan President Paul Kagame since its inception in September 2007. Eliane is a member of the Impact Advisory Board of the Global Alliance for a Sustainable Planet; the Expert Consultation Group on the Post

COVID-19 Implications on Collaborative Governance of Genomics Research, Innovation, and Genetic Diversity; and the African Development Bank's Expert Global Community of Practice on COVID-19 Response Strategies in Africa. She is a member of the newly created Genomic Solution Supervisory Board. Eliane is a member of the Global Crop Diversity Trust Executive Board; the China Council for International Cooperation on Environment and Development (CCICED) Special Policy Study on Post 2020: Global Biodiversity Conservation; and the Ducere Global faculty. Through McGill, Eliane has facilitated leadership experience modules in The Duke of Edinburgh's Emerging Leaders' Dialogues program. She has been a final paper advisor through six cohorts of the International Masters for Health Leadership pioneered by Professor Henry Mintzberg. She is a former member of WWF International's Board of Trustees. She was a facilitator in the International Health Leadership Development Programme (IHLDP) commissioned by the Kenya Red Cross and the International HIV/AIDS Alliance offered by Lancaster University's Management School. She has taught leadership in the International Parliamentary Executive Education program run by McGill University (in English) and by Université Laval (in French) in conjunction with the World Bank Institute. She has also facilitated the UNAIDS Leadership Programme for Women at the United Nation System Staff College.Eliane is a fellow of the African Academy of Sciences. She was the Deputy Executive Director for Programs at Global Open Data in Agriculture and Nutrition (GODAN). Eliane founded C.L.E.A.R. International Development Inc., a consulting group harnessing global networks for sustainable systems development. She was the principal investigator on a Gates Grand Challenges Phase I grant looking at Innovations in Feedback and Accountability Systems for Agricultural Development. She is a past member of FemStep, a research network highlighting rural girls' and women's perspectives for engendering poverty reduction strategies in Rwanda, South Africa, Tanzania, DR Congo and Ethiopia using arts based methodologies. Previously, she was an expert consultant for the non-profit group: The Innovation Partnership (TIP). Eliane was the project manager and an investigator on a Gates Foundation Grand Challenges in Global Health phase project led by Professor Timothy Geary, the Director of McGill's Institute of Parasitology from 2009 to 2014. As a result of this work, she has been a reviewer for the Grand Challenges Canada Stars in Global Health program since 2012. Eliane chaired the International Advisory Board of the African Institute of Biomedical Science and Technology.Prior to going back to academia, she was a scientific director in a Montreal-based biotechnology company for five years in charge of molecular diagnostic and bioinformatics discovery programs. This work led Eliane to undertake consulting work in Haiti and in Africa related to sustainable climate resilient economic growth. She is a contributor to the 2012 released book *The Transforming Leader: New Approaches to Leadership for the 21st Century* by Berrett-Koehler Publishers. Eliane chaired the 15th ILA Annual Global Conference bringing close to a thousand leadership scholars and practitioners to Montreal in 2013. She has been a board member of ILA since. In 2014, Eliane gave a TEDx talk on Reimaging the World from Scarcity to Prosperity. She is a Founding Signatory of the Fuji Declaration that was launched in Japan in May 2015. Eliane contributed a book chapter with Dr. John Baugher to the 2015 ebook *Becoming a Better Leader*. In December 2015, she contributed a piece to the Leading Thoughts section of the *New York Times* in Education on Leadership online platform on *Leadership lessons from experiences of innovation, trauma and grief*. She contributed to the 2016 book *Creative Social Change: Leadership for a Healthy World*.In 2016, Eliane co-led a scoping exercise on Current and Future Science Leadership Development Needs for the Alliance for Accelerating Excellence in Science in Africa. In 2018, she co-facilitated the second Global Women in Science Leadership Workshop in Rwanda with support from the Canadian Institute for Advanced Research. Eliane is a member of the Advisory Boards of ShEquity and Orango Investment Corporation.

Ijeoma Vincent-Akpu , Associate Professor of Ecotoxicology and Hydrobiology, holds a BSc Biology from the University of Lagos and MSc and PhD Hydrobiology and Fisheries from the University of Port Harcourt, Nigeria. She completed her doctorate degree as FGN scholar in 2007, before joining the university as academic staff. She proceeded for post-doctoral research at the University of Stirling, UK, in 2011/2012; international courses in Climate Change Adaptation in Agriculture and Natural Resources Management organized by Wageningen UR in Makerere University, Uganda, and another one in Marine Data and Sample Management and Quality Assurance/Quality Control Procedures in Chemistry/Radiochemistry in Tunisia.Her areas of interest include environmental impact assessment, ecosystem valuation, ecotoxicology and climate change adaptation. She has over 20 years of extensive investigations on fates and effects of pollutants in Niger Delta wetland of Nigeria and Cape Coast of Ghana. Her current research under National Research fund is on the use of water hyacinth as safe biofertilizer that can be used to improve crop yield and remediate crude oil impacted soil.Ijeoma grew up in a commercial city where everyone was into business and educating women was not a priority for families. However, her dad trained her and her other two sisters amidst mockery and disdain from his family. Her dad was her first Mathematics teacher and roused her interest for science which has been fun all through.Ijeoma has received several academic awards and grants, such as Commonwealth Academic Fellowship in 2011/2012; Netherlands Fellowship in 2013; International Atomic Energy Agency (IAEA) Fellowship (2014); SHARE-ACP Mobility Scheme in 2015/2016 as a visiting lecturer to the University of Cape Coast in Ghana; Techwomen Programme in USA (2017); and CV Raman Fellowship as a visiting researcher in ICAR-Central Institute of Brackishwater Aquaculture, Chennai, India, in 2018. She was the only African representative to the World Conference on Indigenous people (Aashukan) in Waskaganish, Quebec, and a research visit to International Fisheries Center at the University of British Columbia at Vancouver, Canada.She was the Acting Head, Department of Animal and Environmental Biology from 2018 to 2020 and Research Associate of Federal

Government of Nigeria-International Atomic Energy Agency (IAEA) Technical Cooperation Projects and Center for Marine Pollution Monitoring and Seafood Safety, University of Port Harcourt, since 2012 till date.Dr. Vincent-Akpu is actively involved in Environmental and Social Impact Assessment (ESIA) studies and training of different non-governmental organizations and over 20 oil impacted communities in Niger Delta under Intervention Framework for Mainstreaming Environmental Rights for Sustainable Livelihood and Development in the Niger Delta, and was the environment expert on Integrated Environmental, Health, Human Right and Gender (EHRG) baseline studies of Ogoni clean up in Niger Delta. She is a resource person for Centre for Environment, Human Rights and Development (CEHRD) and served as expert in review of draft ESIA reports of many projects for Federal Ministry of Environment, Nigeria.In addition to teaching and mentoring at the university and other programmes, she volunteers to capacity building and training of local communities in Nigeria in alternative skills to improve their livelihood and facilitates two community based groups in Niger Delta. Dr. Vincent-Akpu is a mentor and member of several professional bodies such as Co-chair of Agriculture, Forestry and Fisheries Section of International Association for Impact Assessment (IAIA), Secretary of Association for Environmental Impact Assessment of Nigeria (AEIAN), West African Society of Toxicology (WASOT), Zoological Society of Nigeria, British Ecological Society, Association for the Sciences of Limnology and Oceanography, Fisheries Society of Nigeria, Organization for Women in Science for the Developing World (OWSD), Association of Nigerian Women Academic Doctors and others.She has over 70 research publications, several books, book chapters and technical paper presentations at conferences and workshops in Nigeria, Ghana, Portugal, South Africa, Mexico, Switzerland, Italy, the UK, the USA, Canada and India.

Chapter 1
Introduction

Gloria Ukalina Obuzor

Science by Women: A STEM Career Roadmap was inspired by STEMM women's experiences, challenges, and accomplishments in their career and the overwhelmingly resounding success of the monthly Seminar Series initiated by Eucharia Oluchi Nwaichi – led Administration of the Organization for Women in Science for the Developing World (OWSD) University of Port Harcourt (UNIPORT) Branch. Between June 2019 and May 2021, OWSD UNIPORT has hosted 27 paper presentations from diverse fields and geographical locations and hopes to beyond publications on their website (owsd.net) reach out to more audience using this book medium.

The beauty of this 12-section book is that it is written by women scientists of various backgrounds, levels, and stages of their careers as they serve as mentors and mentees to each other and others. These authors have been part of these activities during their STEMM Roadmap.

"Science Education: A Veritable Tool for Development," which explained that science education is the unique field that brings content and process aspects of science to nontraditional scientists, has made a significant positive impact on the world around us. The fields of science, technology, and education hold a paramount place in the modern world, but there are not enough workers globally entering the science, technology, engineering, mathematics, and medicine (STEMM) professions. The chapter made a call for more effort in bringing science content and processes to nonexperts as it is capable of exciting more children into STEMM fields and recruiting more advocates for science-enabling policies.

G. U. Obuzor (✉)
Department of Pure & Industrial Chemistry, University of Port Harcourt,
Port Harcourt, Rivers State, Nigeria
e-mail: Gloria.obuzor@uniport.edu.ng

© The Author(s), under exclusive license to Springer Nature
Switzerland AG 2022
E. O. Nwaichi (ed.), *Science by Women*, Women in Engineering and Science,
https://doi.org/10.1007/978-3-030-83032-8_1

"Taking Research Outcomes to Target Beneficiaries: Research Uptake, Meaning and Benefits" described research uptake simply to include all activities carried out to facilitate the mainstreaming of research outputs/findings into policy and practice and also that the goal of every research is to generate new knowledge or add to existing knowledge hence impact sustainable development.

"Understanding and Utilizing Research Collaborations to Enhance Performance and Visibility of Women Scientists" explained that collaboration in academic research involves partnership between/among academics with regard to a specific project geared towards achievement of beneficial outcomes for all parties concerned. The collaborations can be intra-/interdisciplinary, intra-/interinstitutional, and international.

"Archiving: A Useful Tool for Science and Scientists." An archive is described as an organized collection of the noncurrent records of the activities of a business, government, organization, institution, or other corporate body, or the personal papers of one or more individuals (which is where our personal research activities fall), families, or groups, retained permanently (or for a designated or indeterminate period of time) by their originator or successor for their permanent historical, informational, or monetary value, usually in a repository managed and maintained by a trained archivist. **The added value of archival work is that it captures and preserves what is not published**.

"Wearing Our Gender Lens in Research Design and Development" explained that gender is a crosscutting sociocultural variable and a concept that deals with roles and relationships between men and women that are determined by social, cultural, religious, ethnic, economic, and political factor, and not biology. These gender roles and relationships are a key determinant of the distribution of resources and responsibilities and the power relations between men and women. Gender in research looks beyond the observed roles played by males and females, but explores the societal systems, structures, and power relations that enforce them, that is, the societal values and norms of a given time in the history of a community. However, **to wear a gender lens** means integrating a gender perspective during the planning, implementation, and evaluation of our research which also connotes ensuring that our research has a gender-responsive content, that is, our research will not only identify gender issues (biases) but also proffer solutions to them.

"Impostor Syndrome with Women in Science." "Impostor syndrome" was described as when high-achieving women tend to believe that they were not competent and that they were overevaluated by others. Tips for harnessing impostor syndrome are offered to be ask the difficult questions, find support and be supportive of others, fail forward, wear many hats, stop apologizing, and embrace self-worth.

"The Beauty of Research Data in an Information-Driven World." Research data which was defined as recorded facts or statistics generated and collected for processing and interpretation in a bid to produce original research results can exist in many forms such as text, number, image, audio, video, and script. Organizing research data electronically as on a computer system for easy access and management gives rise to a database. Based on the technique applied to data generation, four main types of research data can be envisaged, such as observational, experimental, simulation, and compiled data. Some helpful tips that can be applied when generating research data were also given.

"Women in Academia: Developing Self-Confidence and Assertiveness." Confidence is defined as the quality of being certain of your abilities or of having trust in people, plans, or the future (it is the key to success, peace of mind, and well-being), while assertiveness is defined as the ability to express your opinions positively and with confidence. It is a key skill that can help you to better manage yourself, people, and situations. The biggest barrier to self-confidence is the "I" factor. The belief "I" am not a confident person, "I" am certain I cannot do this, "I" will not be able to do this, "I" am afraid, or "I" am not sure I can do this. Confront your fears and ask what you are so afraid of. When it is broken down, it may be something that can be dealt with. However, there is the need to step out of one's comfort zone in order to succeed. The negative "I-Factor" threat has to be challenged. Techniques that can help develop an individual assertiveness are given as be your change agent; develop confidence; emotional control; communicate assertively, effectively, and audibly; welcome others' opinions; evaluate progress; and accept that failures or setbacks are inevitable.

"The Secret to Being an Influencer as a Science Leader." **Traits of an influencer are know your values, have confidence, be a maximizer, and communicate effectively. Best practices of a successful influencer are given as** listen to people and create a trusting and compassionate environment; share a clear vision, knowledge, and are inclusive; act as a positive role model for others; give options and latitude, allowing for calculated risks; take responsibility and ownership; inspire and elevate your team members and the whole organization; focus on solutions and avoid blaming others; demonstrate resilience in facing adversity; and celebrate efforts, resilience, and success.

"Ethics in Science Through the Lens of COVID-19 Pandemic" discusses ethics in science which was applied in the situation of COVID-19 that was finally upgraded to a pandemic on March 11, 2020, by the World Health Organization (WHO) has spurred the scientific world into action from availability of the genetic sequence of the virus SARS-CoV-2 which led to a flurry in diagnostics through case managements, information management, and issuance of public health guidelines and public safety measures. Science and its practitioners have tried to live up to its goals of systematic observation, analyses, and logical conclusions. It depended on evidence and was not swayed by conflicts or political shades and persuasions. COVID-19 provided the perfect picture of the flux in scientific research and the painstaking efforts at observations, trying to understand the facts and proffering possible conclusions. Looking through the lens of the pandemic, science seems to have tried to maintain ethical standards even under extreme pressure. Guidelines were put in place to slow the rate of transmission and to give health systems a window of operation without being overwhelmed. There were missteps by scientists in the early days, some of which went against the tenets of ethics that saw even published works retracted flip-flops in mode of transmission but the tenacity to perform a duty to society at lower risks with integrity and honesty while protecting the confidentiality was not in doubt. It shone through. Science must not forget that it only uncovers what is already in existence and must look to nature to understand its laws and become a cooperating link, a voice of truth.

"Time and Resource Optimization for Career Advancement by Women Scientists from Resource-Poor Settings" gave helpful tips to optimize time and included patronage of time management applications such as Microsoft Sticky Notes, setting annual/half-year/quarterly/weekly goals and plan of activities, and utilizing the time and calendar on laptop, desktop, mobile phone, or iPad to set daily reminders. Useful tips like identifying and utilizing free resources, particularly online type to increase your research(er) visibility and collaborative opportunities – personal website/blog builder (e.g., WordPress – see https://tosogbanmu.wordpress.com/) and research profile pages (Scopus, Google Scholar, ResearchGate, and so on) – identifying and working with collaborators with facilities both within and outside your university or country, and collaborating on platforms especially ResearchGate could be great, and developing innovative research that can be conducted with available resources was suggested as a useful tip for optimizing resources.

Happy reading!

Chapter 2
Science Education: A Veritable Tool for Development

Eucharia Oluchi Nwaichi, Ejikeme Obed Ugwoha, and Rosemary Nkemdilim Ogu

2.1 Introduction

Science education covers the teaching and learning of science by non-scientists (Hurd, 1991a, b; Jegstad & Sinnes, 2015) which could include pupils and students in primary and secondary schools (Plate 2.1) and even members of the general public (Plate 2.2). Kids feel proud to become scientists after science education events like Science Festival, Science in the Pub, Street Science, Soapbox Science, etc. Science education grooms curiosity about the world and enhances scientific thinking (Gilbert, 2015) and could identify the qualitative factors that deter women from pursuing careers in STEMM and promote much sought-after diversity. It could bring and keep more women and girls into science, technology, engineering, mathematics and medicine (STEMM) careers (Plate 2.3).

The field of science education involves science content, science process (the scientific method), some social science and some teaching pedagogy. The standards for science education provide expectations for the development of understanding

E. O. Nwaichi (✉)
Department of Biochemistry, University of Port Harcourt,
Port Harcourt, Rivers State, Nigeria

Exchange and Linkage Programmes Unit, University of Port Harcourt,
Port Harcourt, Rivers State, Nigeria
e-mail: eucharia.nwaichi@uniport.edu.ng

E. O. Ugwoha
Department of Environmental Engineering, University of Port Harcourt,
Port Harcourt, Rivers State, Nigeria

R. N. Ogu
Department of Obstetrics and Gynaecology, University of Port Harcourt,
Port Harcourt, Rivers State, Nigeria
e-mail: rosemary.ogu@uniport.edu.ng

Plate 2.1 Next Einstein Forum Fellow, Eucharia O. Nwaichi, and Next Einstein Forum Ambassador, Stephen Manchishi, running science education activity in Zambia. Selected primary and high school students were exposed to coding, and they could create cartoons and similar programmes

for students through the entire course of their kindergarten education and beyond. The traditional subjects included in the standards are physical, life, earth, space and human sciences.

2.2 Content of Science to Non-traditional Scientists

Fenichel and Schweingruber (2010) argued that ascertaining what learning looks like, getting a recipe to measure it and a blueprint to guarantee that people of all ages, from different cultures and settings, have a helpful learning experience should be an all-important consideration for practitioners in informal science backgrounds such as herbaria, museums, libraries, repositories, post-school programmes, science and technology parks, media outfits, aquariums, sanctuaries, wildlife parks and conservatories.

Delivering content of science to non-traditional scientists is complex and requires science leaders who are both task- and relationship-oriented. Gratton and Erickson (2007) posited that relationship-oriented leadership style would be most appropriate

Plate 2.2 Eucharia Nwaichi and her research team at the Institute of Agrophysics, Lublin, Poland, participated at Science Festival in Lublin and came second. To the general public, this science education activity provided an informal and a less-tensed platform for public learning, inspiration and scientific debate

in complex teams, given that team members are more likely to share knowledge in an environment of trust and goodwill. To effectively deliver on science education, objectives must be clear, assigned tasks must be specific and communicated, and monitoring and feedback must be provided. Communication approach (Fig. 2.1) making great considerations for content, human element (from the crusader), structure and packaging of intended message gives good outcomes. STEMM crusaders should strive for increased trust among non-science audience as Nwaichi and Abbey (2015) reported heightened performance and norming frictions between research networks and community members when trust was established. They suggested development of message content and structure, delivery style and presence (Fig. 2.1) to effectively communicate goal, being mindful of entry and appropriate language register in an informal setting.

Fenichel and Schweingruber (2010) posited that the standards for science education provide expectations for the development of understanding to the learners that may include heterogeneous public, children, college students, etc. The traditional science subjects included in the standards are physical, life, earth, space and human sciences. Life science has taught that a chick pecks its way out of the egg, a fingerling fights to get out of the mother fish's belly but a human baby needs a push to get

Plate 2.3 As a UNESCO-L'Oréal Fellow, Eucharia O. Nwaichi worked with Host, Magdalena Frac, to excite more youth (especially girls) into science. Stakeholders like the principal of selected schools in Poland were engaged for sustainability of project outcomes

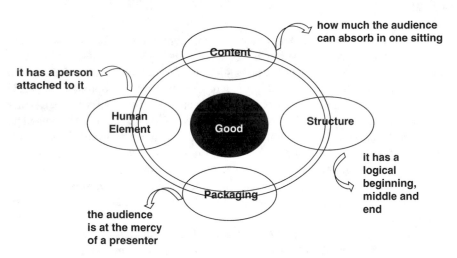

Fig. 2.1 Scientific communication approach. (Nwaichi & Abbey, 2015)

out of the womb. This push in development is the job of science education! A clarion call goes out to learn from these traditional sciences for effective development: break the laws sometimes (physical sciences); innovation comes only by doing 'thinking outside the box' stuff, humans naturally live push life science), encouragement like the science prize inspires more work, protect nature for it is an amazing thing to do (earth science); sustainability, give people some space to innovate (space sciences); spaces are needed to examine performance, the solution to our challenges lie within (human sciences); untapped potentials can be strategically unleashed.

2.3 The Concept of Development

A multitude of meanings has been attached to the idea of development. The term is complex, contested, ambiguous and elusive. However, in the simplest terms, development can be defined as bringing about change that allows people to achieve their human potential. An important point to emphasize is that development is a political term. It has a range of meanings that depend on the context in which the term is used, and it may also be used to reflect and to justify a variety of different agenda held by different people or organizations. This point has important implications for the task of understanding sustainable development, because much of the confusion about the meaning of the term 'sustainable development' arises because people hold very different ideas about the meaning of development. Another important point is that development is a process rather than an outcome. It is dynamic in that it involves a change from one state or condition to another. Ideally, such a change is a positive one – an improvement of some sort. Furthermore, development is often regarded as something that is done by one group (such as a development agency) to another (such as rural farmers in a developing country). Again, this demonstrates that development is a political process, because it raises questions about who has the power to do what to whom.

Development is not simply about the interactions between human groups; it also involves the natural environment (i.e. transforms the environment). So, from another point of view, development is about the conversion of natural resources into cultural resources. This conversion has taken place throughout the history of human societies, although the process has generally increased in pace and complexity with time. Also, development brings about economic growth. From this point of view, development means an increase in the size or pace of the economy such that more products and services are produced. Conventionally, a common assumption has been that if an economy generates more products and services, then humans will enjoy a higher standard of living. The aim of many conventional approaches to development has been to increase the size of the economy in order to increase the output of products and services.

According to Shah (2017), development means improvement in country's economic and social conditions. More specifically, it refers to improvements in a way of managing an area's natural and human resources in order to create wealth and improve people's lives. Dudley Seers while elaborating on the meaning of development suggests that while there can be value judgements on what is development and what is not, it should be a universally acceptable aim of development to make for conditions that lead to a realization of the potentials of human personality (Shah, 2017). Among other conditions, Seers outlined education as what can make for achievement of the aim of development. Empowerment of people takes development much ahead of simply combating or ameliorating poverty. In this sense, development seeks to restore or enhance basic human capabilities and freedoms and enables people to be the agents of their own development. Two major contemporary concerns that require focus in any development initiative are that of human security and sustainability. Fulfilment of basic needs of mankind should be the true objective of development and achievements that either do not contribute to this goal or even disrupt this basic requirement must not be pursued as a development goal.

Additionally, development is the desire and ability to use what is available to continuously advance the quality of life and liberate people from the circle of poverty. It is also tantamount with self-reliance which requires the ability to learn how to advance one's well-being without recourse to others. It involves the ability to act and apply knowledge to improve the knowledge of the process of development and of knowledge itself. Development is linked with scientific and technological progress, modernization, industrialization, electronic and biological revolution, material advancement, the emergence of nuclear energy, new knowledge about man and the universe. It means urbanization, socio-cultural transformation, mass literacy, employment opportunities and the emergence of specialized and independent occupational roles. It includes full growth and expansion of the education, industries, agriculture, social, religious and cultural institutions. The ultimate aim of development must be to bring about sustained improvement in the well-being of the individual and bestow benefits to all self-reliance and mobilization of domestic resources, the transformation of the structure of rural production, the development of small-scale industries and the acquisition of technological and scientific skills. It has been noted that the major factor responsible for the wide gap in the level of development between the so-called developed and developing nations is the level of development of pure and applied science in these nations.

2.4   Sustainable Development

Sustainable development is the organizing principle for meeting human development goals while simultaneously sustaining the ability of natural systems to provide the natural resources and ecosystem services on which the economy and society depend. The desired result is a state of society where living conditions and resources are used to continue to meet human needs without undermining the integrity and stability of the natural system. Sustainable development can be defined as development that meets the needs of the present without compromising the ability of future generations to meet their own needs. According to Kates et al. (2016), the use of this definition has led many to see sustainable development as having a major focus on intergenerational equity. Although the brief definition does not explicitly mention the environment or development, the subsequent sentences, while rarely quoted, are clear. On development, it is stated that human needs are basic and essential; that economic growth – but also equity to share resources with the poor – is required to sustain them; and that equity is encouraged by effective citizen participation. On the environment, the concept of sustainable development does imply limits – not absolute limits but limitations imposed by the present state of technology and social organization on environmental resources and by the ability of the biosphere to absorb the effects of human activities.

Sustainable development is rooted in earlier ideas about sustainable forest management and environmental concerns. As the concept of sustainable development developed, it has shifted its focus more towards the economic development, social development and environmental protection for future generations. It has been suggested that the term sustainability should be viewed as humanity's target goal of human-ecosystem equilibrium, while sustainable development refers to the holistic approach and temporal processes that lead us to the endpoint of sustainability (Shaker, 2015).

In September 2015, the General Assembly adopted the 2030 Agenda for Sustainable Development that includes 17 Sustainable Development Goals (SDGs). Building on the principle of 'leaving no one behind', the new Agenda emphasizes a holistic approach to achieving sustainable development for all. The 17 SDGs to transform our world include Goal 1: No Poverty; Goal 2: Zero Hunger; Goal 3: Good Health and Well-Being; Goal 4: Quality Education; Goal 5: Gender Equality; Goal 6: Clean Water and Sanitation; Goal 7: Affordable and Clean Energy; Goal 8: Decent Work and Economic Growth; Goal 9: Industry, Innovation and Infrastructure; Goal 10: Reduced Inequality; Goal 11: Sustainable Cities and Communities; Goal 12: Responsible Consumption and Production; Goal 13: Climate Action; Goal 14: Life Below Water; Goal 15: Life on Land; Goal 16: Peace and Justice Strong Institutions; and Goal 17: Partnerships to Achieve the Goal. Thus, the concept of sustainability has been adapted to address very different challenges, ranging from the planning of sustainable cities to sustainable livelihoods, sustainable agriculture to sustainable fishing, and the efforts to develop common corporate standards in the UN Global Compact and in the World Business Council for Sustainable Development.

2.5   Development Indices

Gross National Product Index

In the history of development economics, gross national product (GNP) has been thought of as a key indicator in measuring development of a nation. Prior to the 1970s, economic development was evaluated in terms of the GNP and per capita income, which stood alone as the ultimate standard of national progress and prosperity. However, over the years, researchers have found that the GNP is not a single indicator that can be used to measure development. Numerous efforts have been done to create other composite indicators that could serve as complements or alternatives to the traditional measure.

A major breakthrough in the thinking about development came between 1995 and 1999 which led to a re-definition of the development process from one that focuses solely on economic growth to one in which the fruits of economic growth benefit the population in terms of higher literacy rates and education levels, better health and nutrition, higher levels of social cohesion and social skills, and more equality. In a similar line of thinking, the United Nations Development Programme (UNDP) has developed the Human Development Index (HDI), which goes beyond narrow monetary income definitions of development.

Consequently, a new view of development has emerged which is described as a multidimensional process involving major structural changes in social attitudes and national institutions as well as the acceleration of economic growth, the reduction of inequality and the eradication of poverty. Simply stated, development refers to the process of improving the quality of life of all human lives. A major effort in this direction was the development of a composite 'Physical Quality of Life Index' (PQLI). This index was based on a country's life expectancy, infant mortality rate and literacy rate. Overall, good development measurement requires researchers to develop indicators which take into account economic, social, political, environmental and technological concerns.

Many studies have been conducted to identify the indicators that represent the development level of a country. In the meantime, some researchers have come up with a variety of development indices to rank the countries according to their national performance. To construct comprehensive development indices, economic, human, social and environmental concerns and other related representative indicators must be carefully selected. The existing development indices are systematically classified into three main categories of development frameworks, namely human development, social development and sustainable development (Aziz et al., 2015). Indices categorized under human development mainly focus on developing human potentials to the optimum level, and their scopes are mainly for the individual. Indices under the social development category are more all-encompassing – focusing on the good of the greater society, with scopes that go beyond the individual. Meanwhile, indices classified under sustainable development focus on a more multidimensional way where economic, social and environmental dimensions are being considered and integrated.

Human Development Framework

The definition of human development and its aspects are very important as it will affect the choices of performance indicators. The human development concept is broader than other people-oriented approaches to development. It represents a multidimensional and holistic approach to development because it encompasses all aspects of well-being. Human development has been defined by United Nations Development Programme (UNDP) as the process of enlarging people's choices and improving human capabilities (the range of things that they can do or be in life), freedoms, guaranteed human rights and self-respect so they can live a long and healthy life, access to education and a decent standard of living, participate in their community and the decisions that affect their lives. It is expected that the existing indices and indicators used to measure human development will vary based on how they define human development and its dimensions. Some scholars only focused on the physical aspects of human beings in measuring human development, and some others also included the spiritual concerns of human being in their measurement. Since the 1960s, a broader measure of human well-being combining suitable indicators has been created by many scholars and development agencies. The efforts to construct a comprehensive index as a measurement of human development are still active. The discussions on this issue have led to the development of some significant indices to measure human development. The indices are presented as follows.

A. Physical Quality of Life Index (PQLI)

The Physical Quality of Life Index (PQLI) was created as another measurement of human development based on the most basic needs of the people. It is important to note that three main physical indicators, namely life expectancy at the age of 1, infant mortality and adult literacy, were combined to construct the PQLI and used for a cross-country comparison. The index enables researchers to rank countries not by incomes but by the performance of a country in meeting their people's basic needs. Initially, PQLI is developed to study the effect of US aid or assistance given to developing countries. To combine the variables, the scaling procedure which transforms the variables with the lowest value put at 0 and the highest value at 100 was used and a simple average of the three transformed variables was taken to arrive at the required PQLI.

PQLI is a simple and easily computed composite index and can be used to calculate changes in countries over time as well as to measure ethnic, regional, gender and rural-urban differences. PQLI became instantaneously popular because the selected indicators conform with the logical understanding of human development. However, PQLI has not been used for regional comparison and rather focuses on a cross-country comparison. One limitation of the PQLI is that the three indicators were insufficient to capture the quality of life. It fails to take into account problems associated with basic needs like nutrition, health, sanitation, housing, etc. Therefore, the PQLI is inadequate to comprehensively and accurately portray the real level of development of a society. Another limitation of this index is that it measures comprehensive development by only considering the physical aspects of life. By doing so, this index does not include freedom, justice, security and other intangible elements that are important in the overall concept of human development, and, more importantly, only measures how well societies satisfy certain specific life-serving social characteristics.

B. *Human Development Index (HDI)*

It has been correctly recognized that development is much more than just the expansion of income and wealth. Development should emphasize economic growth as a means and not the end of development and therefore should consider health, education, standard of living, human rights, political freedom and self-respect as more important concerns of human development. In 1990, the Human Development Report (HDR) developed a composite index, the Human Development Index (HDI), on the basis of three basic dimensions of human development – to lead a long and healthy life, to acquire knowledge and to have access to resources needed for a decent standard of living. This index has since been equated to human development as it has become one of the important alternatives to the traditional one-dimensional measure of development, the GDP. The HDI contains four variables to represent the three dimensions – life expectancy at birth, to represent the dimension of a long, healthy life; adult literacy rate and combined enrolment rate at the primary, secondary and tertiary levels, to represent the knowledge dimension; and real GDP per capita, to serve as a proxy for the resources needed for a decent standard of living. HDI is similar to PQLI in terms of the indicators but differs on the inclusion of income level in HDI and exclusion of the same from PQLI. HDI attempts to rank all countries on a scale of 0 (lowest human development) to 1 (highest human development) based on three goals or end products of development which are longevity, knowledge and standard of living. HDI is known as one of the most ambitious attempts to systematically and comprehensively analyse the comparative status of socio-economic development in both developing and developed nations. HDI works better than PQLI as a measure of development because it represents both the physical and financial attributes of development. Moreover, HDI is better than other development indices because it effectively facilitates the evaluation of the progress of countries which allows inter-country comparison and inter-temporal comparisons of living levels. This is because HDI uses data that are available in most countries which allow for the widespread international comparisons. The composition of HDI appears to benefit various social policies because the government can specifically find an associated cost or effort required to directly improve the three indicators of HDI. HDI is not short of criticisms, although it has been reported as one of the significant indices to measure countries' performance on the dimension of human development. The small number of indicators in HDI somehow impedes it to successfully capture various aspects of development, thus making it unable to respond better to social problems like corruption. HDI is also argued to be a reductionist measure as it incorporates just a subset of possible human choices and leaves out many aspects of life that are of fundamental importance. Along the same line, HDI has ignored the gender inequality aspect in a society to represent the development of a country. The index also overlooked two important dimensions of human development, which are environment and equity. The exclusion of ecological considerations and equity as indicators of development inhibits the accurate representation of the realities of the world. Furthermore, focusing exclusively on national performance and ranking does not accurately portray development from a global perspective. HDI has been criticized to be an incomplete measure of human development and painted a distorted picture of the world. While HDI carries useful information about a country's current development, it ignores the future level of development as the index used an off-count of past efforts rather than the estimation of the present efforts or prediction of the future. Overall, the HDI has been criticized for not successfully capturing the richness and breadth of the concept of human development. Furthermore, the use of the equal weighted sums of each indicator in the HDI is also an issue. On one hand, the equal-weighted sums of each indicator is a limitation to effectively measure the level ofdevelopment, while on the other hand, it improves the index's goodness of fit given the added complexity of usingassumptions based on unequal weights. From the analysis, it can be concluded that HDI is basically devised as a summary, not a comprehensive measure of human development. Therefore, it is recommended that HDI be refined to be more comprehensive and reflect more aspects of human development and inequalities within a country.

C. *Gender-Related Development Index (GDI) and Gender Empowerment Measure (GEM)*

Besides HDI, other composite measures, a gender related-index, namely Gender-Related Development Index (GDI) and Gender Empowerment Measure (GEM), have been created. Both GDI and GEM were created to include gender inequality issues in human development. GDI takes into account gender inequality in its overall assessment of aggregate human development in a country. GDI measures in the same dimension as HDI, discounting them for gender inequality. This means GDI should be interpreted as HDI discounted for gender disparities in its three components and should not be interpreted independently of HDI. Meanwhile, GEM is meant to be interpreted as an index of gender equity in political and economic participation and decision-making as well as power over economic resources. GEM consists of three indicators which are focusing on empowerment dimension. The selected indicators are male and female shares of parliamentary seats; male and female shares of administrative, professional, technical and managerial positions; and power over economic resources. Since the introduction of GDI and GEM in 1995, several other indicators that directly measure gender inequality have also been constructed such as the Relative Status of Women (RSW) Index, the Standardized Index of Gender Equality (SIGE) and the Gender Equality Index (GEI). These indices were developed due to the shortcomings and misinterpretation of GDI and GEM in many reports and academic writings. Researchers are looking for indicators which directly measure gender inequality. GDI and GEM are both known as rarely used indices which receive minimal attention and have not been highlighted in the international press. This is because of their limited information and empirical value added. In addition, GDI has always been misunderstood by most studies as a direct measure of gender inequality, therefore leading to the misinterpretation and misuse of the index. This shows that the computation of GDI is confusing and vague for people to understand the idea of this index. Besides that, GDI and GEM were also criticized because they do not adequately reflect gender inequality dimensions neither in developing countries nor in developed countries. However, both of these indices have an advantage compared to other gender equality indicators in terms of the separation of dimensions of basic capabilities (GDI) and empowerment (GEM). It has been suggested that it is preferable to separate these two dimensions because different countries may have gender equality in basic capabilities but look very different in the dimension of empowerment and vice versa.

D. *Meaning in Life Index (MILI)*

Taking spirituality of human beings into account, Meaning in Life Index (MILI) was developed and used to see the relationship of MILI with personality and religious behaviours and beliefs among UK undergraduate students. The MILI is an index that measures the extent to which individuals believe their lives have meaning, not just on life quality or life satisfaction. It aims to establish any association that exists between MILI and Eysenck's personality factors of extraversion, neuroticism, psychoticism and social conformity, as well as some religious variables such as church membership and frequency of church attendance. The MILI aspires to cover personality and religious dimension as an extension to the Purpose in Life (PIL) Index. This instrument has gained adequate face validity, internal consistency and scale reliability which allows it for other future research. However, the MILI scores are greatly dependent on who the person is or in other words the personality type. A person's personality has a big influence on how life is considered as having meaning. In addition, data on religiosity showed no association with extrinsic or quest religiosity and a significant but weak relationship between the MILI scores and intrinsic religiosity.

Social Development Framework

Social development can be defined in two ways. It can refer to improvement in the welfare and quality of life of individuals, or changes in societies which make development more equitable and inclusive for all members of a society. These two definitions are based on the meaning of social which refers to people's welfare and to relationships (between individuals and groups within a society). Since 1954, there have been substantial efforts on the areas of social concern and the representative indicators to describe development. The efforts to promote social development in the Western industrial countries can be traced back in the 1970s when social

workers experienced in development work first sought to popularize social development ideas in the United States and elsewhere. Ever since, social development has
become more widely known in these countries. But it is largely as a result of the
World Summit on Social Development that social development has become more
widely known in the Global North. Organized by the United Nations in Copenhagen
in 1995, the Summit called the Copenhagen Declaration on Social Development
addressed a number of pressing global concerns ranging from poverty and unem-
ployment to ethnic conflict and gender oppression. It was agreed, among other
things (Danjuma and Ikpe, 2019), to create a framework for development dedicated
to the eradication of poverty and to increase the resources spent on education and
health. In addition, they pledged to support development that is people-centred and
participatory; that takes account of non-discriminatory and gender sensitivity; that
promotes accountability and transparency in government; and that builds the capacity of all development actors, including the state, the private sector and civil society.
It was also affirmed that economic and social goals are inextricably linked and that
both economic and social factors contribute to sustainable development.

A. *Social Development Index (SDI)*

Social Development Index (SDI) was initiated in 1989 to measure countries'
social development. Multiple indicators have been used to construct this index.
Initially, SDI was created with as many as 13 physical variables to represent social
development across 40 countries. These selected indicators represent urbanization
and industrialization, health conditions, nutritional level, level of education and
social communication dimensions. However, in 2008, SDI was reintroduced with
only 10 physical variables representing various areas of social concern across 102
countries including 21 Organisation for Economic Co-operation and Development
(OECD) countries and socialist countries like China. This index captures a large
number of social indicators to represent more areas of social concern and is associated with an objective method of deriving weights for combining multiple physical
indicators. However, although SDI includes a large number of social indicators to
represent the level of development, the economic condition is being ignored as no
financial variable is included. This is one of the limitations of SDI in presenting a
more holistic view of the development.

B. *Human Poverty Index (HPI)*

Between 1997 and 2009, another composite measure, namely the Human Poverty
Index (HPI), which is an index of human deprivation and a non-income-based measure of human poverty was created. The index values and the rank of countries show
how the intensity of poverty varies across countries. The index recognizes that poverty is multidimensional and that poverty measures based on the income criterion do
not capture deprivation of many kinds. Human poverty is more than income poverty
as it denies people's choices and opportunities for living a tolerable life. The HPI for
developing countries incorporates three types of deprivation as important dimensions of poverty – in survival, in education and knowledge and in economic provisioning. Survival deprivation is measured by the percentage of people (in a given

country) not expected to survive to the age of 40 years; meanwhile, deprivation in education and knowledge is measured by the adult literacy rate. Deprivation in economic provisioning is computed as the mean of three variables: percentage of population without access to safe water, population without access to health services and malnutrition among children less than 5 years of age. The HPI is then obtained as the cube root of the average of cubes of the three above components of deprivation. The HPI can be used in at least three ways – as a tool of advocacy, as a planning tool for identifying areas of concentrated poverty within a country and as a research tool. This composite index has several advantages in determining the social state of development in terms of poverty level for each country. HPI gives a real picture of poverty level in a country since it moves away from income poverty measures to relative deprivation measures and successfully reflects more basic opportunities and choices in terms of survival, education and health. Furthermore, although there is no universal agreement that HPI can identify the causes of poverty, it can illuminate the different dimensions of poverty which policymakers have to address. In addition, like all such indices, the HPI summarizes information especially the extent of poverty along several dimensions. However, it has no obvious merit as a summary measure, particularly in relation to simpler, more easily understood, indices such as the simple mean. Another limitation of the HPI is about the particular choice of variables for describing and quantifying deprivation and about the reliability of the data actually used. There are so many related variables being excluded from the index which are the deprivation in terms of food, clothing and shelter as applicable to the whole population, deprivation of gainful employment, deprivation of basic human rights including equality before law and justice, etc. It has been emphasized that the concept of human poverty is actually larger than the HPI. This is because some important dimensions are difficult to quantify, or data do not exist. Examples of such dimensions include political freedom, personal security and exclusion.

C. *Multidimensional Poverty Index (MPI)*

In 2010, HPI was supplanted by Multidimensional Poverty Index (MPI). The index was developed by Oxford Poverty & Human Development Initiative (OPHI) and the United Nations Development Programme (UNDP). The MPI constitutes a set of poverty measures which can be used to create a comprehensive picture of people living in poverty. The index offers a valuable complement to traditional income-based poverty measures by considering multiple deprivations at the household level. The index identifies deprivations across the same three dimensions as the HDI with ten indicators; two represent health (malnutrition and child mortality), two are educational achievements (years of schooling and school enrolment), and six aim to capture standard of living (access to electricity, drinking water, sanitation, flooring, cooking fuel and basic assets like a radio or bicycle). The three broad categories (health, education and living standards) are weighted equally (one-third each) to form the composite index which shows the number of people who are multidimensionally poor (suffering deprivations in 33% of weighted indicators) and the number of deprivations with which poor households typically contend. The MPI relies on three main databases that are publicly available and comparable for most developing countries: the Demographic and Health Survey (DHS), the Multiple Indicator Cluster Survey (MICS) and the World Health Survey (WHS). There are some advantages of the MPI compared to the HPI. The index is able to capture the severe deprivations that each person faces at the same time and can reflect both the incidence of multidimensional deprivation and its intensity – how many deprivations people experience at the same time. Thus, this addresses the shortcoming of the HPI which could not identify specific individuals, households or larger groups of people as jointly deprived as it used country averages to reflect aggregate deprivations in health, education and standard of living. In addition, the MPI can be broken down by indicator to show how the composition of multidimensional poverty changes for different regions, ethnic groups, urban and rural location as well as other key household and community characteristics. This is why MPI is described as a high-resolution lens on poverty as it can be used as an analytical tool to identify the most prevailing deprivations. Besides, the methodology of MPI shows aspects in which the poor are deprived and help to reveal the interconnections among those deprivations. This enables policymakers to target resources and design policies more effectively. This is especially useful where the MPI reveals areas or groups characterized by severe deprivation. However, the MPI also has several drawbacks. First, the indicators included in this index are from different elements because the data are not available for all dimensions. Some indicators are based on outputs (such as years of schooling) and others based on inputs (such as cooking fuel). Second, in order to be considered the multidimensional poor, the MPI stated that households must be deprived in at least six standard of living indicators or in three standard of living indicators and one health or education indicator. However, data availability for all indicators is questionable. Therefore, careful judgements were needed to address missing data in some cases. Third, while the MPI goes well beyond a headcount to include the intensity of poverty experienced, it does not measure inequality among the poor, although decompositions by group can be used to reveal group-based inequalities. Finally, the estimates are based on publicly available data which limits direct cross-country comparability. These drawbacks are mainly due to data constraints. With these drawbacks, it is expected that this index will evolve over time just like the other development indices.

D. *Corruption Perception Index (CPI)*

The Corruption Perceptions Index (CPI), created by a non-governmental organization, the Berlin-based Transparency International (TI), and first released in 1995, has been designed to provide a more systematic and extensive snapshot of corruption within countries. These perceptions enhance our understanding of real levels of corruption from one country to another. The CPI combines a number of different indicators into one composite index to measure corruption. This means that the CPI is a homogeneous index in the sense that all the components upon which it is based seek to measure the same thing. The CPI is based on data collected over a number of years prior to release of the index. As the calculation of CPI is the combination of data sources, the index that results from it is highly reliable because the probability of misrepresenting a country is lowered. Other than that, by using CPI, human behaviour and attitude towards investment decision, political participation and any other activities can be predicted. From this prediction, government and other related institutions may plan for further action to counter all the consequences and possibilities. However, CPI does not reflect the actual corruption incidence experienced by a country, and it does not explain the characteristic of a country that may affect the calculation of the index. Moreover, corruption is an issue which has a broad concept of discussion, and it can occur in many ways. Thus, CPI does not define specifically what perception on corruption is being measured. Other than that, CPI also may be biased as the index may be on the perception of the people of government or the people on the opposition side.

Sustainable Development Framework

There has been a significant research effort to define and operationalize measures of development. In practice, measures of development tend to concentrate explicitly only on economic and social dimensions and neglect the aspect of the environment. However, since the Earth Summit in Rio de Janeiro in 1992, the role of social and environmental indicators has become the focus of much attention. Rio's Agenda 21 commits all 178 signatory countries to expand their national accounts by including environmental costs, benefits and values. This worldwide interdisciplinary effort to integrate economic, social and environmental dimensions in development measurement is aiming to put all countries on its path towards sustainable development. Sustainable development has been defined in many ways, but the most frequently quoted definition is from the Brundtland Report in 1987 where development is said to be sustainable if it 'meets the needs of the present without compromising the ability of future generations to meet their own needs'. This explains sustainable development as a multidimensional concept of development where economic, social and environmental dimensions are being considered and integrated. In addition, sustainable development considers the long-term perspectives of the socio-economic system, to ensure that improvements occurring in the short term will not be detrimental to the future status or development potential of the system. Sustainable development has also been defined as minimizing the use of exhaustible resources such as energy, water, land and air, or at least ensuring that revenues obtained from them are used to create a constant flow of income across generations, and making an appropriate use of renewable resources. Also, three main components of sustainability have been noted which are environmental, social and economic sustainability. Environmental sustainability is described as natural capital remaining intact which means the functions of the environment should not be degraded. Meanwhile, social sustainability requires the cohesion of society and its ability to work towards common goals be maintained and at the same time all basic human needs be met. Economic sustainability means the country is financially feasible when development moves towards social and environmental sustainability. Therefore, in 2005, the German Council for Sustainable Development agreed that the concept of sustainability has to be extended beyond environmental concerns, to include social and economic sustainability. The well-being of these three areas is so intertwined that it is difficult to neatly separate them. This means that people today have to leave the future generations an intact ecological, social and economic system. These definitions show that meeting the needs of the future depends on how well we balance social, economic and environmental objectives or needs when making decisions today. Nationally, there are some well-developed SDI programmes such as Sustainable Seattle 1993, and some have been given the lead by existing State of the Environment (SOE) reporting programmes such as SOE Canada 1991. However, it has been noted that despite the considerable attention devoted to SDIs in several years since 1992, no set has emerged with universal appeal, and new SDI sets experience difficulty in gaining wide acceptance. In addition, there are many arguments provoked on the usefulness of the indicators selected to promote sustainable development. Initially, there were three most common sustainable development indicators, namely economic, social and environmental, used in the literature. It has to be realized that most of the existing indicators are the aggregated single index where only one variable is reported. These essentially identical indicators involve the estimation of a range of economic, social and environmental benefits in monetary terms but with a different name. In 1989, the Index of Sustainable Economic Welfare (ISEW) was introduced. This index takes into account commuting costs and the costs of accidents, water, air and noise pollution, loss of farmlands and wetlands and others. Then, in less than 10 years, in 1995, the Genuine Progress Indicator (GPI) was introduced. Following that, in 1999, the Sustainable Net Benefit Index (SNBI) was introduced. However, it was observed that these three measures are far from ideal and might cause confusion since they have identical indices but go by different names. This aggregated single index, such as the ISEW, is not widely used, although the index is receiving quite considerable academic attention for several years and has been applied to the United States, the United Kingdom and Scotland. The most widely accepted move towards sustainable development measurement is the development of methods for 'green accounting' which includes ecological and resource stock valuation in the system of national accounts. In Green Reporting, all indicators involved measure real-world results. However, it is argued that 'green' GNP, like all economic-based measures, can never be an adequate measure of sustainable development due to the problems with evaluating common goods that exist outside the market place and due to problems in elucidating social equity. Generally, all these existing single aggregated indices of sustainable development are not likely to be adequate if used alone because they are difficult to be applied at regional and local scales due to patchy data availability. Also, these indicators are not user friendly as they are not readily understood by the laymen. These single aggregated indices may well communicate changes in sustainable development but are unlikely to be effective in identifying the changes that are required to promote sustainable development at the local level. So, a set of simpler indicators is required to better promote sustainability. This set of SDIs is expected to complement the use of the single aggregated index and is essential to promote sustainable development at all levels. However, it has been established that indicators produced by one group of specific local authorities are often found to be unsatisfactory to another. In order to produce a common set of SDIs and implement it on a larger scale, a preliminary analysis of composite indicators of sustainable development using Spearman's rank correlation was proposed. There are numerous single and composite indicators of sustainability used which include environmental, social, economic and sustainable development dimensions. They were the Direct Material Consumption Index (DMC), well-being (WB), Ecological Well-Being Index (EWB), Environmental Sustainability Index (ESI), ecological footprint (EF), CO_2 ecological footprint (EFCO2), Human Development Index (HDI), Dashboard of Sustainability (DS-SDI), Dashboard of Sustainability Environmental Sector (DSEnv), Geobiosphere Load (GBL), Gross Domestic Product (GDP), Happiness Indicator and Quality of Life (QoL). In summary, it is indeed very difficult and challenging to identify a core set of SDIs common to all localities and to produce a genuine composite index of sustainable development which addresses global sustainability concerns.

2.6   Nexus Between Education and Development

Science Education and National Development

It is a global knowledge that there is a positive relationship between education and economic, political and cultural development. According to Chabbott and Ramirez (2000), two rationales played a major role in buttressing confidence in the relationship between education and development. The first constructs education as an investment in human capital, which will increase the productivity of labour and contribute to economic growth and development at the societal level. This rationale is closely tied to global norms about science, progress, material well-being and economic development. The second general rationale constructs education as a human right, imagining education as the prime mechanism for human beings to better themselves and to participate fully in the economy, politics and culture of their societies. This rationale is tied to notions of justice, equality and individual human rights (Igbaji et al., 2017).

Education, like other forms of investment in human capital, can contribute to economic development and raise the incomes of the poor just as much as investment in physical capital, such as transport, communications, power or irrigation. Education supports the growth of civil society, democracy, political stability and citizens' rights. Education can contribute to the development of human rights, human development, human capital and social cohesion. In today's knowledge-driven economies, access to quality education and the chances for development are two sides of the same coin. That is why targets must be set for secondary education while improving quality and learning outcomes at all levels. That is what the sustainable development goal on education aims to do; hence, governments should work with parent and teacher associations, as well as the private sector and civil society organizations, to find the best and most constructive ways to improve the quality of education.

A sound educational system is known to be the pivotal for sustainable develop-ment of every nation. Generally, science is considered as the process through which knowledge is arranged in an organized pattern. That is, science could be described as the structure and behaviour of the physical and natural world and society, espe-cially through observation and experience. Science education emerged as an applied field of education saddled with the responsibility of disseminating scientific skills and knowledge. In other words, science education is concerned with the sharing of scientific knowledge with people not traditionally considered part of the scientific community. It must be emphasized that science education transforms the typical teacher-centred classroom lecture into a discovery and problem-solving arena.

The practical impact of scientific research enables the emergence of science poli-cies and influences the scientific enterprise by prioritizing the development of com-mercial products, healthcare, public infrastructure and environmental fortification. There is currently a dearth of human resource for science education (Eilks, 2015; Jonna, 2020). The world must put in place concerted efforts to increase the number of young boys and girls studying science. Science education is a veritable tool for pushing out these concerns and in turn bringing development. The process of sci-ence education encourages creativity and originality, which demands the active engagement of students in identifying problems and looking for solutions. Hence, teaching and learning of science education address issues that are typical to local environment and expose students to national issues in other environments around the world, thereby producing students that are globally inclined to thinking. Accordingly, the goals of science education are to (a) cultivate inquiring, knowing and rational mind for the conduct of a good life, (b) produce scientists for national development, (c) service studies in engineering/technology and the cause of techno-logical development and (d) provide knowledge and understanding of the complex-ity of the physical world, the forms and the conduct of life.

Science education is desired to meet the needs of industries and citizens as well as satisfy the practical needs of the society (Miswaru and Sadiyya, 2017). It is directed towards acquiring critical thinking and exploration. It is a process of teach-ing or training especially in school to improve one's knowledge about the environ-ment and to develop one's skill of systematic inquiry as well as natural attitudinal characteristics.

Science education is germane to the scientific and technological advancement of any nation.

This is because science education comprises the comprehensive study of proven scientific concepts and principles. McCarthy (2017) describes progress made with science education in China and the resulting bumper harvest of STEMM graduates. He reported 40% completion of a degree in STEM subject in China, more than twice the share in American third level institutions. He also documented increasing importance of workers with STEM qualifications to global prosperity, and unsur-prisingly, China is leading the way (Fig. 2.2).According to McCarthy (2017), China had 4.7 million new STEM graduates in 2016, India had 2.6 million, while the

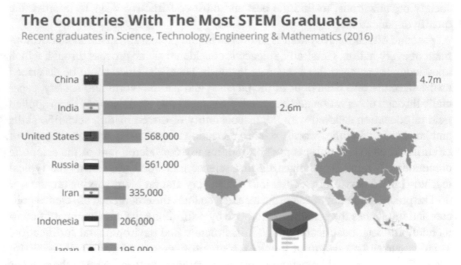

Fig. 2.2 Countries with the most STEMM graduates. (McCarthy, 2017)

United States had 568,000. Recent STEM graduates in 2016 as presented at a World Economic Forum and reported in the work of McCarthy (2017).

Science education identifies natural phenomena appropriate to a child's interest and skills. This implies that science education equips teachers, learners and society with knowledge, skills, equipment and freedom to perform noble tasks useful for improving socio-economic standards. Thus, the goal of science education is to produce a sufficient number and diversity of skilled and motivated future scientists, engineers and other science-based professionals.

However, it has been reported that if the aim of science education has to be accomplished in any country, then there is the need to promote effective teaching of science subjects in the schools right from primary through secondary to tertiary levels (Onwukwe & Agommuoh, 2016). Also, it has been opined that part of the reason why many developing countries may not attain their Development Goals is curriculum-based. In addition, it has been explained that the problem with science education in most developing countries is a lack of a good curriculum and must be fixed for the aim of science education to be realized (Onwukwe & Agommuoh, 2016).

Application of Science Education (Engineering) and Development

Engineering plays a crucial role in supporting the growth and development of the economy of a nation as well as in improving the quality of life for mankind. Thus, there is a vital link between a country's engineering capacity and its economic development. Engineering covers numerous types of activity. Engineers make things, make things work and make things work better as well as design solutions to the world's problems and help build the future. Engineering has been defined by the Royal Academy of Engineering as the 'creative application of scientific principles', principles that are put into practice to invent, design, build, maintain and improve structures, machines, devices, systems, materials and processes. This definition is broad and intended to account for the fact that the scope of engineering is continually evolving.

Engineers are responsible for some of the most important advances in biomedicine, and they have played a key role in building the infrastructures around mankind, ranging from roads to utility networks. Engineers also play a role in the processing of foods and the development of new materials to be used in manufacturing. With millions of people living in poverty and without sufficient food or sanitation, engineering continues to have a key role to play in helping countries to progress across the world.

Economic theory suggests that growth in the economy, which is the only means of increasing the prosperity of a country, depends on the quantities of the factors of production employed (labour and capital) and the efficiency with which those quantities are utilized. Growth is sustained by increasing the amounts of labour and/or capital that are used and by increasing the efficiency with which they are used individually and in combination to produce output. Countries in the economic development phase must focus on improving the efficiency of utilization of labour and capital. Economic development is vital in creating the conditions necessary to achieve long-run national growth.

It must be noted that as each additional unit of the factors of production (labour and capital) is added, the resulting amount of additional output tends to diminish. Only increases in the level of technological progress can offset this decline in growth that occurs as diminishing returns to labour and capital set in. Growth over the long run is sustained by increasing the efficiency with which these factors are combined to produce output, a process known as total factor productivity (TFP). Improvements in TFP are driven by a number of variables including the depth and breadth of technical knowledge – as reflected in things like standards, patents and licences (permissions to use, produce or resell). Other drivers include the quality of education, the average number of years of education among the wider population or investment in research and development.

Therefore, economic development, while difficult to precisely define, results from investment in the generation of new ideas through innovation and the creation of new goods and services, the transfer of knowledge and the development of viable infrastructure. Examples of economic development include the creation of infrastructure, not just roads and bridges, but also digital and communications infrastructure, and the creation of knowledge through education and training, which can be utilized by businesses to create new goods and services.

By investing in infrastructures, such as transport, bridges, dams, communication, waste management, water supply and sanitation as well as energy and digital infrastructure, countries can raise their productivity and enhance other economic variables. By having a well-developed transport and communications infrastructure, for example, countries are better able to get goods and services to market and move workers to jobs. A strong communications network allows a rapid and free flow of information, helping to ensure businesses can communicate and make timely decisions. All of these infrastructure projects require engineers, products of science education.

Ways in Which Application of Science Education (Engineering) Contributes to Economic Development

Engineering is an extensive field that can contribute to economic development through several channels. By investing in infrastructures, such as transport, bridges, communication, waste management, dams, water supply and sanitation, energy and digital infrastructure, nations can increase their productivity and improve other economic variables. By having a well-developed transport and communications infrastructure, for example, nations are better able to move goods and services to market and workers to workplaces. A strong communications network allows a swift and unrestricted flow of information, helping to ensure businesses can communicate and make timely decisions. In other words, no nation can have an economy without engineering (application of scientific knowledge). This is because engineering plays a vital role in the production of goods and services, through creating new knowledge and ensuring there is the capacity in place to produce and move goods and services (such as infrastructure, transportation networks and logistical arrangements). Engineering can also help address challenges that will help countries to meet the United Nations Sustainable Development Goals aimed at ending poverty, fighting inequality and injustice and tackling climate change by 2030.

Engineers help countries by developing infrastructure that provides basic services such as energy; water and food security; transport and infrastructure; communication; and access to education and healthcare. Linked to these goals, engineering should also have a positive impact on factors such as life expectancy that over time can be expected to aid economic development through improvements to productivity, which in turn results in increased GDP. Figure 2.3 illustrates the strong relationship between the quality of infrastructure in a country and the level of economic development achieved across the world. This supports the assertion that engineering contributes to economic development as it has a key role to play in ensuring countries have a strong infrastructure.

2.7   Current and Future Role of the Application of Science (Engineering)

According to the United Nations Educational, Scientific and Cultural Organization (UNESCO), engineering has been, and will continue to be, confronted with designing systems that facilitate education and healthcare, improve quality of life and

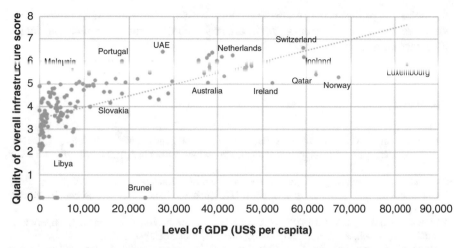

Fig. 2.3 Relationship between the quality of infrastructure in a country and the level of economic development. (Royal Academy of Engineering, 2016)
(Quality of overall infrastructure score: 1 = extremely underdeveloped, 7 = well developed)

assist to eliminate global poverty. It considers that the development of technological approaches that can help prevent or mitigate hostile acts, reduce the impact of natural disasters and motivate humans to reduce their use of the earth's valuable resources will be key challenges for engineering in the coming years. Alongside these, it is expected that engineering will continue to play a key role in helping to avert environmental crises as well as helping to reduce poverty via the provision of community infrastructure. Engineering already plays an important role in managing and conserving resources, from water to food, energy and materials. For example, engineering skills have been essential in ensuring the development of systems relating to water and wastewater treatment. Given that some parts of the world still lack access to water, engineering skills will remain essential to ensure universal access to clean water and sanitation. Engineering has also been extensively involved in finding solutions to reducing carbon emissions alongside ensuring increased portions of the world's population have access to sustainable power. Engineering's role in this area is likely to continue to be important in the coming years, especially as in 2015 it was estimated that 2.8 billion people still did not have access to modern energy services and that over 1.1 billion people were without electricity (Royal Academy of Engineering, 2016). In addition, with the global population expected to grow to 9.7 billion by 2050, engineering will become increasingly important in ensuring future food security (Royal Academy of Engineering, 2016), for example, by ensuring that there are sustainable food production systems in place that maintain ecosystems and by helping to improve land and soil quality. Over and above these growth areas, UNESCO envisages new challenges for engineering across four key areas: materials, energy, information and systems and bioengineering. Each of these fields will require engineers across a range of disciplines to ensure future innovations and success. Therefore, having sufficient numbers of engineering graduates and professionals focusing on engineering for development in these areas will be essential both now and in the future and should sit unmistakably at the centre of science education.

References

Aziz, S. A., Amin, R. M., Yusof, S. A., Haneef, M. A., Mohamed, M. O., & Oziev, G. (2015). A critical analysis of development indices. *Australian Journal of Sustainable Business and Society, 1*(1), 37–53.

Chabbott, C., & Ramirez, F. O. (2000). Development and Education. In M. T. Hallinan (Ed.), *Handbook of the Sociology of Education* (Handbooks of Sociology and Social Research). Springer.

Danjuma, S. G., & Ikpe, A. (2019). Science education and sustainable development in Nigeria: An analytic approach. *IOSR Journal of Humanities and Social Science, 24*(6), 29–34.

Fenichel, M., & Schweingruber, H. A. (2010). *National Research Council. Surrounded by Science in Informal Environments*. The National Academies Press. https://doi.org/10.17226/12614. ISBN 978-0-309-13674-7.

Gratton, L., & Erickson, T. J. (2007). *Eight ways to build collaborative teams*, Retrieved January 12, 2021, https://hbr.org/2007/11/eight-ways-to-build-collaborative-teams

Gilbert, J. K. (2015). *International Encyclopedia of the social & behavioral sciences* (2nd ed.). Elsevier.

Hurd, P. D. (1991a). Closing the educational gaps between science, technology, and society. *Theory into Practice., 30*(4), 251–259.

Hurd, P. D. (1991b). Issues to Linking Research to Science Teaching. *Science Education, 75*(6), 723–732.

Igbaji, C., Miswaru, B., & Sadiyya, A. S. (2017). Science education and Nigeria national development effort: The missing link. *International Journal of Education and Evaluation, 3*(5), 46–56.

Eilks, I. (2015). Science education and education for sustainable development – Justifications, models, practices and perspectives. *Eurasia Journal of Mathematics, Science & Technology Education, 11*(1), 149–158.

Jegstad, K. M., & Sinnes, A. T. (2015). Chemistry teaching for the future: A model for secondary chemistry education for sustainable development. *International Journal of Science Education., 37*(4), 655–683.

Jonna, T. (2020). *Why is science education important?* Retrieved from https://jyunity.fi/en/thinkers/why-is-science-education-important/ on 22 Apr 2021.

Kates, R. W., Parris, T. M., & Leiserowitz, A. A. (2016). *Environment: Science and Policy for Sustainable Development* (pp. 1–13). Editorials. Taylor & Francis Group.

McCarthy, N. (2017). *Recent graduates in STEMM*. https://www.industryweek.com/talent/article/21998889/the-countries-with-the-most-stem-graduates Retrieved 23 April 2021.

Miswaru, B., & Sadiyya, A. S. (2017). Science education and Nigeria national development effort: The missing link. *International Journal of Education and Evaluation, 3*(5), 46–56.

Nwaichi, E. O., & Abbey, B. W. (2015). Essentials of scientific communication in a productive research. *Journal of Scientific Research and Reports, 7*(7), 525–531.

Onwukwe, E. O., & Agommuoh, P. C. (2016). Science and Technology Education: A veritable tool for peace, conflict resolution and national development. *IOSR Journal of Research & Method in Education (IOSR-JRME)*. e-ISSN: 2320–7388, p-ISSN: 2320–737X, *6*(4), 37–42.

Royal Academy of Engineering. (2016). *Engineering and economic growth: A global view*. A report by Cebr for the Royal Academy of Engineering.

Shah, S. (2017). *Development: Meaning and concept of development.* Sociology Discussion. Retrieved from https://www.sociologydiscussion.com/society/development-meaning-and-concept-of-development/688# on 20 May 2021.

Shaker, R. R. (2015). The spatial distribution of development in Europe and its underlying sustain ability conditions. Applied Geography 63 304.

UNESCO. United Nations Educational, Scientific and Cultural Organisation. Science Education | United Nations Educational, Scientific and Cultural Organization. Retrieved May 18, 2021., from http://www.unesco.org/new/en/natural-sciences/special-themes/science-education/

World Health Organisation WHO. *Diabetes.* Retrieved May 18, 2021., from https://www.who.int/news-room/fact-sheets/detail/diabetes

Chapter 3
Taking Research Outcomes to Target Beneficiaries, Research Uptake, Meaning and Benefits

Clara Chinwoke Ifeanyi-obi

3.1 Introduction

Research cycle does not end at the discovery of new knowledge. It goes further to sharing the knowledge discovered and hence impacting attitude, behaviours, practices and policies. Huge resources are invested in conducting research with the hope that solutions to problems will emanate through these researches. Research is expected to stimulate change through the dissemination and adoption of new knowledge generated. Productivity of industries could be improved through application of research outcomes.

What happens when these research outcomes remain only in the shelves of universities and research institutes? The aim of conducting the research has not been fully achieved and value for the resources invested not obtained. Practices and policies not informed by research evidence hardly meet the felt needs of the people. It is important for researchers to understand the critical need of sharing their research outcomes. The objective of this chapter is to explain the concept of research uptake and the benefits of conducting research uptake both to the target audience and researchers and give a guide to researchers on the process of conducting research uptake. It is believed that understanding the meaning, importance and process of research uptake will stimulate an increased number of research uptake activities carried out among researches.

C. C. Ifeanyi-obi (✉)
Department of Agricultural Economics and Extension, University of Port Harcourt,
Port Harcourt, Nigeria
e-mail: Clara.ifeanyi-obi@uniport.edu.ng

© The Author(s), under exclusive license to Springer Nature
Switzerland AG 2022
E. O. Nwaichi (ed.), *Science by Women*, Women in Engineering and Science,
https://doi.org/10.1007/978-3-030-83032-8_3

31

3.2   Research: Concept and Purpose

Concept

Research is simply an inquiry to better understand a subject matter, its occurrence, reasons behind its occurrence, new knowledge about it and application of these knowledge. Every person not minding the area of specialization is involved in one aspect of research or the other. When you search for answers to questions, try to understand new phenomenon, reasons behind occurrence of new phenomenon, application of new knowledge to existing phenomenon, understand logic behind some mathematical models, you are indulging in research. Research could be described precisely as a search into new knowledge. It involves information search, processing and sharing. Sekeran (2006) in Madukwe and Akinnagbe (2014) defined research as a process of finding solutions to a problem after analysis. In scientific research which is the bedrock of innovations, an organized way or defined methodology is followed in its search to make the outcomes reliable and valid. Scientific research involves a systematic investigation of a defined subject matter with the purpose of generating new knowledge on the case study. It leads to the discovery and invention of innovations; hence, sustainable development anchors on it. Bryman (2012) listed three criteria that could be used to evaluate scientific research as follows:

- Reliability – It checks if the research study is repeated under same conditions, can same results be obtained; that is, are the measures used reliable and consistent.
- Replication – This ensures that sufficient details of a research methodology are reported such that another researcher could repeat the study.
- Validity – This concerns the integrity of conclusions that are generated through a research study. It checks if the measure employed accurately reflect the concept under investigation, the robustness of the relationships between variables, if the research findings can be extrapolated beyond the research context.

Scientific research follows a defined method of search which is dependent on the type of research being conducted or the field of study.

Purpose

Every research is targeted towards achieving some stated objectives. Madukwe and Akinnagbe (2014) stated that the purpose of research is to find answers to questions or solutions to problems through application of specific procedures. Research fulfils three main purposes which include:

(a) Explores information on new area of knowledge. It builds the foundation for more detailed or extensive study on a particular topic.

Fig. 3.1 Research cycle. (Source: Google, https://www.google.com)

(b) Describes and validates the behaviour of a case study using information or data collected from such population.

(c) It goes further to explain the reason or the 'why' for a particular behaviour or phenomenon.

It is important to note that a research lifespan has a cycle which does not terminate at finding answers to research questions; it goes further to ensure that the findings are shared and the desired impact is made. Figure 3.1 is the diagram of a research cycle showing the five stages with the last stage of research as sharing knowledge and making impact.

3.3   Research Uptake: Meaning and Benefits

Research uptake simply put are all activities carried out to facilitate the mainstreaming of research outputs/findings into policy and practice. It involves all the activities that facilitate and contribute to the use of research evidence by policymakers, practitioners and other development actors (DFID, 2016); it explained that research uptake activities aim to support the supply of research through communicating research findings effectively to potential users (not only disseminating findings through a peer-reviewed journal article) and the usage of research outcomes by building capacity and commitment of research users to access, evaluate, synthesize and use research evidence. The goal of every research is to generate new knowledge or add to existing knowledge add to existing knowledge, hence impact sustainable development. Therefore, for any research to meet the ultimate goal, it must initiate change. Most researchers do not understand the importance of ensuring that their research outcomes reach their target audience. When research findings lie idle in bookshelves or at most get published in journals, it cannot be said to have met the expected impact. This is because only a small proportion of the target audience get to read journal publications, hence the need to identify possible ways of mainstreaming research findings into policy and practice.

Furthermore, when researchers engage the public in an effort to mainstream their research findings to policy and practice, they get feedbacks on their project which helps to identify gaps and possible restructuring of future works. This is to say that research uptake activities do not only benefit the public, researchers themselves are also enriched through feedbacks received; hence, uptake activities form basic part of monitoring and evaluation of research projects.

Research cycle is never complete without knowledge sharing and impact; therefore, researchers must ensure that they revolve through the complete research cycle to ensure expected impact is made and maximum value obtained from resources invested in the research activities.

Some donor agencies fund research as a way of contributing to the sustainable development goal of poverty reduction (DFID, 2016). This is done through production of new knowledge and products or technologies that directly improve the lives of poor people. These new knowledge and products will only have the desired impact if it is understood and used to inform decisions. Therefore, conducting research uptake activity ensures intentions/objectives of the donor agencies in research funding are achieved.

Research uptake is an opportunity to engage local stakeholders in research project. REFANI (2016) listed benefits and lessons learnt from local stakeholder engagement in research project to include arousing research project interest in stakeholders, facilitate uptake, establish the project and its partners in the field and build a lasting interest among local partners to track uptake of evidence, even after the project ends.

3.4   Brief Guide on Implementing Research Uptake Activity

When Do We Begin to Plan Our Research Uptake Programme?

Research activities commence with the thinking and planning stage. To successfully mainstream research outputs into policy and practice, research uptake activities must be inculcated into research plan right from the planning stage. This will ensure that all relevant stakeholders are identified and carried along in the course of the research. When stakeholders become part of the research project activities, it stimulates their interest and builds their sense of commitment towards the project; hence, mainstreaming of the research findings becomes a hitch-free one. Bringing in all relevant stakeholders from research inception also helps in making the research activity holistic as all stakeholders have the opportunity to contribute their own ideas and perspective to the research project according to their different experiences.

Defining Objectives and Type of Research Objective Activity to Hold

It is important for researchers to specify the objective of intended research uptake. This will not only guide in ensuring that SMART objectives are listed, it helps researchers in selecting appropriate research uptake activity that will ensure the intended objectives are achieved. The objectives of a research uptake activity mostly depend on the outcomes and recommendations of the research leading to the uptake activity. If the research recommends knowledge building or capacity development for increased uptake of an innovation, then a training workshop will be the ideal uptake activity (Figure 3.2); if it demands a review or embedment of an issue into policy or development of supportive policies, then a policy discourse or dialogue will be suitable; if massive awareness is recommended, then strategic campaign may be suitable. The nature of your research determines what your uptake activities will be. There could also be a combination of different uptake activities to ensure a particular objective is maximally achieved.

Examples of research uptake activities include:

- Workshops/trainings (with researchers, uptake practitioners, research users/beneficiaries, policymakers)
- Conferences to promote exposure to and use of the research by various stakeholders
- Policy forums and symposia that bring together researchers, practitioners and policymakers for discussions around the research

Fig. 3.2 Research uptake process

- Developing publicity material; infographics, training manuals, newsletters, bulletin, policy brief, success stories
- Advisory services to stakeholders
- Production of documentaries

Identifying Stakeholders in Research Uptake Programme (Stakeholder Mapping)

A stakeholder is an individual, group of people/organization who is interested and impacted by the outcome of a research project. They also include all those who are capable of having a major effect on the research project. There are two categories of stakeholders in any research uptake programme, namely primary and secondary stakeholders.

Example of List of Key Stakeholders for a Policy Discourse on Mainstreaming Climate Change Adaptation Options into Policy and Practice in Southern Nigeria(*Programme convened by the author with support from the Department for International Development (DFID) under the Climate Impact Research Capacity and Leadership Enhancement Programme (CIRCLE)*
- National Root Crops Research Institute (NRCRI) management staff
- Representatives from the Ministry of Agriculture
- Management staff of Agricultural Development Programme (ADP)
- Representatives of State Environmental Protection Agency (SEPA)
- CIRCLE fellows, mentors, supervisors and ISP leads in universities concerned, namely Michael Okpara University of Agriculture Umudike, University of Port Harcourt, University of Ibadan
- Professional and community-based association leaders
- Leaders of farmers cooperatives and associations
- Key community leaders
- Leaders of civil society associations
- Researchers in climate change and agriculture
- Key farmers

Primary stakeholders are direct beneficiaries of research outcomes/interventions, for example, funding agencies, institutions/organizations and users of the information (farmers in agriculture sector), while secondary stakeholders include all indirect beneficiaries of research outcomes/interventions: government agencies (regulatory bodies, etc.) and policymakers. Who constitutes primary and secondary stakeholders in any uptake programme depends on the objectives of the programme. The ability of the researcher to identify and involve the key stakeholders in any research uptake activity determines to a large extent the achievement of the uptake objectives. To ensure all key stakeholders in a research uptake programme are listed, it is advisable for researchers to make a list of specific objectives that will culminate in achievement of the overall uptake activity. Then list the stakeholders according to the specific objectives. This will help to ensure that no key stakeholder is omitted.

A Practical Guide on Mapping Stakeholders in Research Uptake: How to do it1— Once you have identified the desired research uptake outcome for your study, as a group identify and list all the actors that may affect the policy or practice change, both positively and negatively. Focus your attention on the most relevant or well-known actors, such as policymakers (e.g. government departments or government officials), implementers (e.g. non-governmental organisations), practitioners (e.g. doctors, nurses), researchers or individuals. Remember to think about both internal and external actors, as well as those at global, regional, national and district levels, as appropriate.2— Plot these actors onto these stakeholders on the stakeholders analytical employed according to their level of alignment and interest. This should be based on evidence about their current behaviours. To assist you, think about the following: Alignment: Do they agree with our approach? Do they agree with our assumptions? Do they want to do the same things we think need to be done? Are they thinking what we are thinking? Interest: Are they committing time and money to this issue? Do they want something to happen (whether it is for or against what we propose)? Are they going to events on the subject? Are they publicly speaking about this?3— Prioritise the actors you have identified. Consider the influence/power of each and their accessibility. Mark the actors you would like to prioritise with a red circle, as shown in the figure overleaf. Ideally, the actors you prioritise should be both influential and accessible, but it may also be appropriate to focus on non-influential but highly accessible actors.4— Develop a pathway of change for your target audiences, suggesting a trajectory (represented by the arrows) that you expect or hope each actor will follow. For example, do you want some actors to increase their interest, alignment or both?5— Once you have completed this exercise. You can separate your stakeholders into district, national, regional and international levels, if appropriate. **Extracted from Guide to developing and monitoring a research uptake plan. Malaria consortium. Disease control, Better health.**

Developing Research Uptake Message

In developing the message to be communicated during research uptake programme, the language must be tailored to meet the needs, level and understanding of the target audience. When the uptake activity is to be carried out within a local setting, the use of local language in developing the message is very critical to command the interest and increase understanding among the local people.

Care must also be taken to ensure that the content of the message covers the intended objective of the uptake. The message must address the needs of all identified stakeholders in the uptake programme. Where the key stakeholders are highly heterogeneous, the message could be developed separately for each stakeholder to ensure each stakeholder's need is addressed.

The message must be concise and precise to avoid confusion that arises from ambiguity. The researcher must always bear in mind that the audience may not be specialists in the area of research hence must trim down the message in a concise manner to facilitate understanding.

Identifying Effective Communication Channel and Produce Materials

Effective communication is a very important component of a research uptake programme. Once the stakeholders in a research uptake activity are well defined and identified, the process of identifying effective communication channel is simplified. It is important for researchers to have significant knowledge of the stakeholders, particularly their level of knowledge as regards the topic at hand and prevalent communication channels to enable them to choose appropriate channels that will effectively communicate the message. Using a combination of oral, written and visual channels is noted to produce better results as this makes longer impression in the receiver's memory (Rogers, 2003; Wejnert, 2002). Also, the number of persons to reach in an uptake programme is an important factor to be considered in choosing communication channel to be used. Where the objective is to reach a huge number of persons spread over far geographical locations, a mass media method is advisable.

In contemporary society where digital tools have become more accessible and limitations of physical mobility exist due to global issues, social media has gained wide acceptance and usage among both scientific and non-scientific audience. Integrating the use of social media in research uptake communication holds great potentials as it does not only enable researchers to engage a huge number of people effortlessly, it could be a subtle way to overcome bureaucratic delays associated with accessing multinational organizations by tagging their accounts in a post. Some helpful social media platforms in this regard include Twitter, YouTube, LinkedIn, Facebook and Instagram. Other researcher exclusive social networking platforms where researchers can share their outcomes with other researchers include AuthorAID, Academia.edu and ResearchGate.

In producing materials for uptake activity, simplicity and clarity is the watchword. Researchers most often are carried away with their scientific jargons forgetting that their audience in most cases do not understand those terms. Using simple and clear languages facilitate understanding among the audience. Diagrams, pictures and sketches could also be very helpful in helping your audience to understand the topic.

Copyright issues should also be considered in producing materials for uptake. Researchers are advised to use credible publishers to secure the copyrights of the materials they developed during uptake programme.

Hold Team Meeting to Assess Preparedness

Sharing research outcomes cannot be successfully done by just the researcher. Many non-scientific skills are required including the ability to communicate with multi-stakeholders (Fisher et al., 2020).

A preview meeting before the actual uptake event is very important to ensure all important issues are addressed and duties assigned. Test running the equipment to be used during the uptake programme could also be done during this preview meeting. This helps to identify envisage challenges during the programme hence build in strategies to checkmate them. If the planned research uptake activity is a physical event, it is advisable to hold the preview meeting in the proposed venue for the programme where necessary. This will help the team to have on-the-spot assessment of the available facilities and repairs to be done before the programme. Where it is not possible to hold the programme in the proposed venue, a visit to the proposed venue is advised before the proper event. Where resource persons will be needed in the programme, it is important to collect their slides and preview during this meeting to ensure the content is in line with the programme objectives.

3.5   Conclusion

It is important for policymakers to have reliable evidence for policy decisions. This can only be possible if researchers make a conscientious effort to make their research findings available to policymakers. Without the relevant data for policymaking, it will be difficult to have policies that truly address the needs of the people. Therefore, for nations to have informed policy decisions, more visibility of research outputs through uptake programme is needed.

In addition, adoption of innovations and improved technologies developed through research cannot be possible unless stakeholder's knowledge, skill and capacity in the use of the innovation are developed. Conducting uptake programme to build capacity of potential users of innovation is one effective way of facilitating increased innovation uptake.

References

Guide to developing and monitoring a research uptake plan. Malaria consortium. Disease control, Better health. https://www.google.com/search?q=Guide+to+developing+and+monitoring+a+research+uptake+plan&rlz=1C1YTUH_enNG942NG942&oq=Guide+to+developing+and+monitoring+a+research+uptake

Research on Food Assistance for Nutritional Impact (REFANI). (2016). *What does it mean to implement a research uptake strategy? Experiences from REFANI Consortium*

Bryman, A. (2012). *Social Research Methods* (4th ed.). Oxford University Press.

Sekeran. (2006). *Research methods for business. A skill building approach*. Wiley.

Madukwe, M. C., & Akinnagbe, O. M. (2014). Identifying research problems in Agricultural Extension. In M. C. Madukwe (Ed.), *A guide to Research in Agricultural Extension*. Agricultural Society of Nigeria (AESON).

Department for International Development. (2016). *Research uptake: A guide for DFID-funded research programmes*. Department for International Development and Association of Commonwealth Universities. Research uptake in a virtual world. A guide

Fisher, J. R. B., Wood, S. A., Bradford, M. A., & Kelsey, T. R. (2020). Improving scientific impact: How to practice science that influences environmental policy and management. *Conservation Science and Practice., 2*, e210. https://doi.org/10.1111/csp2.210

Rogers, E. (2003). *Diffusion of Innovations, 5th Edition*. Simon and Schuster. ISBN Ụ/Ҳ·ĮĮ·/Ạ·4/·ҔҲ/ꓒ·ꓕ·Ạ

Wejnert, B. (2002). Integrating models of diffusion of innovations: A conceptual framework. *Annual Review of Sociology, 28*, 297–326. https://doi.org/10.1146/annurev.soc.28.110601.141051. JSTOR3069244.S2CID14699184

Chapter 4
Research Collaborations for Enhanced Performance and Visibility of Women Scientists

Chioma Blaise Chikere, Memory Tekere, Beatrice Olutoyin Opeolu, Gertie Arts, Linda Aurelia Ofori, and Ngozi Nma Odu

4.1 Introduction

By definition, research collaboration involves the working together of individuals brought together by varied reasons for a common goal or purpose. In collaboration, there is an interdependence of scholars for reasons that include opportunities to

---

The original version of this chapter was revised. The correction to this chapter is available at https://doi.org/10.1007/978-3-030-83032-8_14

C. B. Chikere (✉)
Department of Microbiology, Faculty of Science, University of Port Harcourt,
Port Harcourt, Nigeria

Department of Environmental Sciences, College of Agriculture and Environmental Sciences (CAES), Florida Science Campus, University of South Africa (UNISA),
Johannesburg, South Africa
e-mail: chioma.chikere@uniport.edu.ng

M. Tekere
Department of Environmental Sciences, College of Agriculture and Environmental Sciences (CAES), Florida Science Campus, University of South Africa (UNISA),
Florida, Johannesburg, South Africa
e-mail: tekerm@unisa.ac.za

B. O. Opeolu
Extended Curriculum Programmes Unit, Faculty of Applied Sciences, Cape Peninsula University of Technology (CPUT), Western Cape, South Africa
e-mail: Opeolub@cput.ac.za

G. Arts
Wageningen Environmental Research, Droevendaalsesteeg 3 - 3A, 6708 PB Wageningen,
Wageningen University and Research, Wageningen, The Netherlands
e-mail: gertie.arts@wur.nl

share knowledge, resources and tasks (Katz & Martini, 1997; Gazni & Didegah, 2011). In sciences, Lee and Bozeman (2005) echoed that nowadays science is becoming more complex, inter-transdisciplinary and costly such that research collaborations offer immense advantages, thus becoming a norm in modern-day research. Individuals, institutions and scientific research organizations cannot underscore the importance of research collaborations. In fact, funding organizations and instruments place a requirement for teamwork and collaboration on individuals as part of their requirements (Quadrio-Curzio et al., 2020). Research studies have clearly demonstrated that productivity, output and quality are enhanced in cases where skills and synergy are applied in a collaboration (Zhang et al. 2020). Collaboration encourages creativity, leverages experience, maximizes resources and promotes quick turnaround time and solutions. Also, collaboration provides for a mix of professional levels, e.g. high ranks and low, as well as experienced and emerging researchers. Also, projects done in a consortium by virtue of high quality and completeness are often published in high-impact factor journals and offer potential for academic-private partnership which can increase an academic's productivity through follow-on publications and innovation. Other benefits include improved reputation and other spins such as tenure, new and large grants, better postgraduate students' throughput and other research accolades (https://www.nap.edu/read/2109/chapter/3) (NAP, 1993). In science and technology, women face many challenges as reported elsewhere (Zhang et al., 2020), and collaborations offer much needed support. Noting that men are more open to collaborate with men and for different strategic reasons, while women are more open across collaborations, woman to woman collaboration work helps bridge some under representation and gender differences in scientific performance (ArauÂjo et al., 2017). A study by Zhang et al. (2020) with research scientist respondents from Chinese universities showed that women performed better in collaborations and their academic performance improved significantly with international collaboration. Collaborations among women are noted to work well and with an appreciation of challenges and space that women function and operate from. Advantages offered by female scientists in collaboration include effective social skills, good collaboration network construction and cooperative relationships (Zhang et al., 2020).

L. A. Ofori
Department of Theoretical and Applied Biology (TAB), Kwame Nkrumah University of Science and Technology, Kumasi, Ghana
e-mail: laandoh.cos@knust.edu.gh

N. N. Odu
Department of Medical Laboratory Sciences, 1 TAP Road, Elelenwo, Off KM 17, Pamo University of Medical Sciences, Port Harcourt, Rivers State, Nigeria
e-mail: nodu@pums.edu.ng

4.2   Purposeful Collaboration and Collaborators

Collaboration must be approached carefully depending on the basis of why one sought out to or is being sought into the collaboration. One has to do his or her homework to know and understand the potential collaborators. It is essential to reach out to people that one can collaborate with, whether physically or online, and talk to people at conferences and scientific events to meet new or potential collaborators. Butts et al. (2019) highlighted that successful collaborations require matched collaborator interests, high-quality project management, mutual benefits, necessary resources and flexibility to adjust to changes. At the same time, poorly designed projects, culture clashes, systems bureaucracy, poor project management, low benefits and poor supportive organizational structures were noted to work towards ineffective collaborations.

Collaboration puts a requirement on one's skills and time; thus, as a collaborator, one has to be invested as expected (Abramo et al., 2017). Also, the nature of collaborators dictates on the collaboration timing. For example, if a funding instrument requires collaboration as one of the funding conditions, it places a requirement on the potential applicant to identify collaborators beforehand. For the process not to be rushed and compromised, oftentimes it works best for one to be in a collaboration arrangement prior to the project. Collaboration, alternatively, can be sought out during an ongoing project if there is a need for certain skills, expertise or equipment that are not at their disposal. In a collaboration, all parties must at least have something to contribute, ranging from exceptional skills, expertise, competence, unique environmental placing to specialised equipment. With technological advancement, distance has become less and less of a factor and barrier in research collaboration (Metcalfe & Blanco, 2021).

4.3   Types of Collaboration

Types of collaborations range from intra-/inter-disciplinary, intra-/inter-institutional to international. Collaborations can stem from individual, institutional, professional, regional and at international levels, and the benefits from the collaboration accrue to all these levels too. Individuals, institutions and professional organizations are paramount in driving regional and international collaborations (Quadrio-Curzio et al., 2020). Successful individual level collaborations seem to be easy to manage and provide better spins out of the relationship when compared to those dictated at higher levels of collaboration organization. Intra-institutional collaboration between colleagues in same department, different departments and faculties allows researchers that are in physical proximity to each other to work together. Regional collaborations often place researchers in a similar cultural setting and more similar research problem context. International collaborations remove the research and skills divide as they cover researchers of different nations to reach out and find synergy in their research, training and development. Gaillard et al. (2014) show that research collaborations between the developed (EU) and developing countries (Latin American and Caribbean (LAC)) provided for a win-win face of collaboration and better supportive schemes.

4.4   Factors to Be Considered Before Initiating Research Collaborations

Successful collaborations take into consideration a number of key factors from the onset. Time, commitment, shared interest/objectives and expected outcomes all have a bearing as to the setting up of collaborations. Time is of essence as there is a need to commit and deliver as per set timelines. Failing to commit time often leads to the breakdown of collaboration and can lead to researcher productivity fall (Abramo et al., 2017). Working committedly with shared interest and objectives is paramount to the deliverable of successful collaboration. According to the participants at the Gender Equity Workshop at Molecular Approaches to Malaria 2020 Conference, held in Lorne, Australia, from 23 to 27 February 2020, various strategies and resources that helped them overcome obstacles during their careers in parasitology include identifying early in their career a diversity-prone working environment, appropriate mentors and a good group of collaborators that seem to be important for women to avoid isolation, stay connected and further their networking skills to boost their careers in science. It was apparent from discussions at the workshop that this is not a straightforward path. Gender equity, diversity and inclusion in STEMM careers are all work in progress. Being diligent, resilient, patient, persistent and remaining aligned with the core principles and values that inspire women to start a career in STEMM are all desirable qualities to successfully navigate this challenging path (Hansen, 2020).

4.5   Role of Scientific Organizations in Fostering and Funding Collaborations

As highlighted, collaborations require time, space and finances; professional/scientific organizations often bring together people with skills and expertise in similar discipline, gender or different socio-economic settings. By offering such platforms and opportunities, organizations like TWAS, OWSD, and Elsevier Foundation bolt opportunities for identification, creation or supporting collaborations (Quadrio-Curzio et al., 2020; Álvarez-Bornstein & Bordons, 2021). Often some collaborations would not occur were it not for the support and requirement placed on applicants to collaborate. Research funding programmes such as those for travel and research exchange allow for funding aligned to specific goals and objectives to be provided for. Because the funding organization has set timelines and deliverables, such institutions can ensure high rate of collaborative success and turnover. Research studies suggest that funding plays a role in promoting high-impact research, decreasing uncited papers and fostering teamwork. Funded research is more complete as it promotes publication in high-impact factor journals, high uptake and highcitation rate of the research papers emanating from robust sponsorship (Álvarez-Bornstein & Bordons, 2021).

4.6 Impact of COVID-19 Pandemic on Research Collaboration in STEMM: Pathways to Survival for Women in Science

COVID-19 has affected lives irrespective of race, gender, age or location. In the wake of this pandemic, research collaborations, especially those externally funded, have been affected in diverse ways – mainly in meeting targets, timelines and road-maps, constant interactions and meetings, among others. Several traditional research work packages and activities were postponed or cancelled to begin again when things normalized. Other proposed research works were suspended to fund COVID-19-related research. It has been documented that there have been wide-spread impacts on research activities. Many young researchers lost grants for the support of small to high-risk studies leading to bigger grants as a result of the impacts of this pandemic (Korbel & Stegle, 2020). Yet in spite of this, creativity and innovations have enabled us to still drive our research agenda.

4.7 Creativity and Innovation in Research Amidst the Pandemic

The wake of COVID-19 revealed the limitations that have existed in collaboration and communication structure and services around the globe in the face of extensive attention to scientific progress and its politicization (Radecki & Schonfeld, 2020). In times past, research took place at few sites, with few persons, data and research ideas were secretive, shared only among a few restricted to the study with only part shared with the scientific community. Currently, the search for better understanding of this novel virus and the quest to develop a cure have increased collaborations within countries and the world (Roehrl et al., 2020). This pandemic has changed the ways scientific research is conducted in most places. Research has been carried out on COVID-19 with partners in several countries not meeting physically but planning and executing high-level research in the midst of this pandemic (Iba et al., 2020; Sylverken et al., 2021). In Africa, women scientists have been at the forefront in such efforts, collaborating with renowned researchers around the globe as well as supporting the fight against the pandemic (International Science Council, 2021). Data sharing is an important activity in the midst of a pandemic especially when little is known about the biology of the organism which seems to be mutating at a fast rate. Sharing of gene sequence data is also vital in bringing to the fore genomic information on the different variants of SARS-CoV-2 (Glăveanu & de Saint 2021).

4.8   Managing the Virtual Space for Optimal Research Output Following Travel Restriction

COVID-19 hit most countries with its associated lockdowns of land, air and sea borders for weeks up to several months. Yet, several researches still continued in spite of the challenge with collaborations at different levels all for a common good. Virtual platforms (Zoom, Google Meet, Microsoft Teams) took the centre stage in the management of research within and between countries just as databases (Scopus, PubMed, ResearchGate) helped with informing the scientific community of ongoing trends (Singh et al., 2020). Social media also became a privileged space for routine and participative creativity through the production and sharing of coronavirus information either for pleasure or for research during and after travel restrictions in most countries (Richard et al., 2020).

4.9   Role of Professional Societies in Creating Beneficial Research Collaboration

Professional Societies – General, Discipline- and Gender-Specific Societies

Many academics use professional societies as vehicles for developing their careers (Ansmann et al., 2014; Bickel, 2007; Xu & Martin, 2011). Professional societies can serve as sources of information, advice and support and can potentially bring junior and senior members of a society together (Ansmann et al., 2014). Besides career building, other potential benefits are transfer of professional norms, gaining young members and promoting long-term commitment to academic careers (Ansmann et al., 2014; Bickel, 2007). Professional networks might be either informal or formal (Xu & Martin, 2011). Informal professional networks have been proven beneficial for job effectiveness and career advancement (Xu & Martin, 2011).

In the scientific arena covered by STEMM, many professional societies are active at the continental and global levels. Examples in the field of STEMM are SETAC (Society of Environmental Toxicology and Chemistry), SOT (Society of Toxicology), FEMS (Federation of European Microbiological Societies), IUMS (International Union of Microbiological Societies) and SRA (Society for Risk Analysis), a bunch of ecological societies just to be named a few examples. Most of these professional societies have a focus on a specific scientific field or a scientific arena.

Gender-specific societies can help women with their challenges and opportunities in science: these societies function as networks of women from which they can learn from each other, share experiences, help each other, share best practices and give the opportunity to young career scientists to develop themselves and build up a network. Just knowing that the challenges we face are also faced by others is very valuable. Others might have overcome certain challenges, which mean nothing is impossible with the right focus and making use of your inner strength. Building networks is of utmost importance in early career development. Combining bottom-up networking with mentoring programmes aids women in career building. Women are as ambitious and intelligent as male counterparts but often lack confidence. Therefore, they need supportive environments and women role models to succeed.

Examples of gender-specific organizations are the Organization for Women in Science for the Developing World (OWSD) and the European Platform of Women Scientists (EPWS). OWSD is an international organization founded in 1987 and based at the offices of The World Academy of Sciences (TWAS) in Trieste, Italy. It is a programme unit of UNESCO. OWSD is the first international forum to unite eminent women scientists from the developing and developed worlds with the objective of strengthening their role in the development process and promoting their representation in scientific and technological leadership. OWSD provides research training, career development and networking opportunities for women scientists throughout the developing world at different stages in their careers. EPWS is an international non-profit organization that represents the needs, concerns, interests and aspirations of more than 12,000 women scientists in Europe and beyond.

Charters are a specific initiative at the level of universities and research institutes to encourage and recognize commitment to advancing gender equality. One of them is the Athena SWAN Charter Award. This charter was launched in 2005 and recognizes commitment to advancing and promoting women's careers in STEMM employment (see https://www.ecu.ac.uk/equality-charters/athena-swan/). The Athena SWAN Charter is a framework which is used across the globe to support and transform gender equality within higher education (HE) and research. The Charter helps institutions to achieve their gender equality objectives.

Benefits Derivable from Active Participation in Professional Activities

Individual scientists can experience a range of benefits in joining professional societies and networks (Xu & Martin, 2011); they can socialize with senior colleagues; enhance professional-based relationships; experience opportunities for peer evaluation and enhance the individual's reputation and visibility in the field; get up to date with the current developments in the field; and receive informal recommendations and valuable job information. Mentoring is one form of an informal professional network. Ansmann et al. (2014) argue that it is beneficial to have a network of mentors for providing career-related support that might range from developing specific skills to feedback and emotional support. Such a network of mentors fits with the view that academic careers are no longer linear processes but a series of learning cycles (Ansmann et al., 2014). There are no doubts about the importance of professional networks; however, research has revealed significant differences between female and male network interactions (Xu & Martin, 2011). Especially in male-dominated STEMM research fields, women experience difficulties to access the networks (Xu & Martin, 2011).

Professional Societies that Women Scientists Can Join to Enhance Their Careers

One of the professional societies to be mentioned here is the Society of Environmental Toxicology and Chemistry (SETAC; setac.org). SETAC has three main principles on which the society is based, namely multidisciplinary approaches to solving environmental problems, multi-stakeholder engagement and science-based objectivity. It is a global society that has a policy with focus on equity, diversity and inclusion. In all its activities, boards, committees and interest groups, SETAC thrives for equal multi-stakeholder representation, geographical representation and gender balance (Nüßer et al., 2019). SETAC is building a culture of diversity and inclusion by the activities of several interest groups (at the North American and European levels). SETAC North America organized a Special Session at SETAC SciCon2, 'Diversity in Environmental Sciences', which focused on the positive impacts on science when equity, diversity and inclusion become part of the scientific process. The SETAC North American board also renewed the SETAC North America membership in the 'Societies Consortium on Sexual Harassment in STEMM', which is an initiative to advance professional and ethical conduct, climate and culture. SETAC North America, along with 130 other professional non-profit organizations in the fields of STEMM, are working together to combat this widespread issue. SETAC is committed to changing inadequate and inequitable policies, practices and customs that affect awards, meetings, expected conduct in SETAC activities and more. With the help of the Consortium, SETAC North America hopes to help eliminate sexual, racial and other intersecting bases of harassment in STEMM and create a truly inclusive community. SETAC has several policies in place (https://www.setac.org/page/policies). In October 2019, the SETAC World Council issued the declaration on research integrity and linked the SETAC Code of Ethics to it. Last year, the SETAC journals adopted a double-blind peer review in an effort to reduce unconscious bias, and SETAC Europe held a session on gender bias at their Helsinki meeting. Within SETAC, SETAC Africa organizes the Africa Women's Event (SAFWE) every 2 years during its scientific congress (Opeolu et al., 2018). The purpose is to create a safe place for women who participate in the SETAC Africa Biennial Conferences to discuss issues and challenges that are specific to women in science.

4.10   Utilization of Research Resources to Enhance Performance and Visibility

Publishing research data: What, Where and with who to publish

Data processing and final dissemination of curated scientific information through academic channels have been very serious issues in the academia and research Institutions. Two decades ago, publication of scientific data was more or less done manually from editing stage to bibliography stage especially in the developing world. However, in recent times with the advent of ICT-enabled platforms and editing managers, publishing of research outputs for researchers has become less cumbersome as so many of the stages of data management, manuscript preparation and proof reading are better done by just 'a click of the mouse' of a computer. Eminent publishers around the world have equally developed value-added software, reference managers, text editing and grammar check guides to this effect. Elsevier, for instance, has robust journal finder tools that assist scientists to get appropriate journals to possibly submit their manuscripts for publication by matching the research topics, abstracts and keywords with available journals in the database as guide. This approach saves time and at the same time gives the researcher the flexibility in choosing the most suitable journal that fits into the findings and import of their research. SETAC also has two notable journals with good impact factors that welcome research outputs as short communication, systematic reviews and full-length scientific papers. These journals published by Wiley in the United Kingdom are Environmental Toxicology and Chemistry and Integrated Environmental Assessment and Management. They provide a forum where scientists and other professionals can exchange ideas, collaborate and develop sustainable strategies to mitigate environmental problems (SETAC, 2021).

Research becomes beneficial and encompassing when there are enabling resources freely available especially to scientists in low and lower-middle income economies to conduct meaningful and impactful studies and investigations as well as channels for information dissemination and sharing. In developing countries, access to low-cost or most preferably free resources has been a very nagging setback in driving sustainable scientific advancements. Universities and research institutions are poorly funded, which in turn hinders notable scientific discoveries, value-added product development and capacity progression that could be good boosters of economic transformation and growth. Interestingly, in recent times, there has been a revolution in information, communication and technology (ICT) facilities leading to an incredible improvement in the availability of information at the fingertips of researchers, provided there is uninterrupted Internet facility. Notable global publishers like Elsevier, Hindawi, SpringerNature, Taylor and Francis and Wiley, to mention a few, have well-developed platforms to assist scientists to be connected to their peers globally just by the click of a mouse. Predatory publication is a major problem confronting scientists in research and development as it erodes the need for credibility, honesty and quality science since the requirement of publication in journals with good impact factors for promotion assessments and the high cost of publication tend to encourage researchers to circumvent these criteria. However, countries like South Africa through their Higher Education Department prescribe journals that they recognize and subsidize for publications by research and academia (Moodley & Gouws, 2020).

Career-Enhancing Web Resources

Scopus (sign up at https://www.scopus.com/) indexes content from 25,000 active titles and 7000 publishers which are thoroughly vetted and selected by an independent review board, and uses a rich underlying metadata architecture to connect people, published ideas and institutions. Using sophisticated tools and analytics, Scopus generates precise citation results, detailed researcher profiles and insights that drive better decisions, actions and outcomes (Scopus, 2021). BrightTALK (https://www.brighttalk.com) webinars provide webcasting technologies in knowledge industries for Elsevier, ScienceDirect, Scopus and other research organizations. These presentations equally give participants networking and collaboration opportunities. Mendeley (sign up at https://www.mendeley.com) equally provides research updates, personal publication statistics and opportunities to create, join and grow collaborative research groups. Researcher Academy (sign up at https://www.researcheracademy.elsevier.com) offers free online modules with certificates developed by global trailblazers that connect participants to notable experts in different fields of science that are well suited for career guidance and advice. One beautiful thing with all Elsevier career advancement platforms is the flexibility in that the researcher just uses one log-in ID (id.elsevier.com) to sign in to several resources, thereby circumventing the challenge of developing several usernames and passwords for access.

Role of Active Researcher Profile Databases in the Projection of Women Scientists' Achievements (African Scientists Directory)

This connects African scientists globally (register through http://africanscientists.africa/); ORCiD (Open Researcher and Contributor ID) uses robust digital infrastructure to enable researchers to share their information and equally connect to their peers on a global scale. Researchers can register at https://www.orcid.org. ORCiD connects through unique/persistent identifiers all active participants in research, scholarship and innovations and their research outputs and contributions across disciplines, borders and time. Researchers can also link their ORCiD profiles to Scopus through easy and user-friendly steps. Other beneficial academic platforms researchers can sign up with for visibility and networking are Google Scholar, Mendeley, ResearchGate and Web of Science (most of which give author's metrics and citation indices).

Registration and Active Presence on Professional Social Media Platforms

Women scientists are encouraged to have a robust social media influence to project their academic profiles. By sharing one's achievements and notable research activities on professional online networks like LinkedIn, Twitter and Facebook, a researcher increases the visibilty and dicoverability of herresearch and professional engagements (preferably following and liking pages of professional bodies and notable science organizations, e.g. United Nations arms, World Bank, granting and funding bodies, connecting with prominent scientists). These are good platforms that are well suited to enhance the career of women scientists.

4.11   Mentoring as a Tool for Career Growth and Development

Globally, women scientists are faced with numerous diverse challenges that limit career achievements and progression. These career-limiting constraints include competition with the boy child for family resources to get education, 'octopus woman burden' status – the multiple traditional roles that compel women (as a daughter, wife, mother, aunt, niece, cook, laundry lady, role model, pastor) to be everything to everyone, workplace challenges (such as bullying and sexual harassment), the pressure to work harder to prove competencies, 'PhD (pull her down) syndrome' versus boys' club, access to resources for personal growth and the need to sacrifice personal career development opportunities, among others. These challenges make mentoring imperative for women scientists to effectively mitigate some of them. This section therefore focuses on roles of mentoring for women scientists, strategies for structured and effective mentoring programmes and some benefits of mentoring.

Importance of Mentoring for Women Scientists

Akyavuz and Asıcı (2021) defined mentoring as 'a sharing process between two individuals – the mentor and the mentee'. The mentoring journey should ideally transcend academic support for the mentee. It also includes social and psychological support when needed (Akyavuz & Asıcı, 2021). The role of effective mentoring strategies cannot be overemphasized for all scientists, but the need is greater for women and other under-represented gender groups. There is significant amount of empirical information on the benefits of mentoring in communities and professional practices (Akyavuz & Asıcı, 2021; Bell & Rosowsky, 2021; Egbctokun & Olofinyehun, 2021).

The goal of mentoring programmes is to guide the mentee to effectively set and achieve career goals. The mentor provides information and/or gives pointers for accessible career development opportunities. Women scientists that have competent mentors often perform better in the various metrics of measurement for success. Guidance and support from senior colleagues motivate mentees' career choices. The tendency to mimic role models pushes mentees to apply research funds, awards, training, travel grants, etc. Women without mentoring opportunities, however, struggle to navigate the professional terrain due to other issues. Structured systematic mentoring for grant-writing skills doubled the potential of a proposal for funding relative to those submitted by individuals that were not guided (Botham et al., 2021; Smith et al., 2021).

Sustainable availability of women scientists requires capacity development across generations. Women leaders are essential players in the empowerment campaigns and programmes of nations. They serve as role models to girls in communities and emerging scientists in workplaces. Capable women scientists flow into the positive cycle of uplifting other women and girls. Mentoring is an essential tool that has proven to enhance career progression and accomplishments. It helps mentees to take beneficial career trajectories and make best career decisions. For a mentoring programme to be successful, the mentee needs to be guided to have a career vision board, mission (goals and strategies), career choices and opportunities, acquisition of hard and soft skills and spirituality for psychosocial strength. The need for mentoring women scientists, therefore, cannot be overemphasized.

Mentoring Strategies for Career Performance Improvement

The decision to mentor and be mentored is a career duty rather than a job responsibility. Most female scientists focus on job descriptions and make life choices that often stifle career growth and development. Mentoring provides both the mentor and mentee the opportunity to learn from each other and grow together along the journey. A successful mentoring programme requires the willingness to engage by both individuals.

The relationship between the mentor and the mentee is important for a successful mentoring programme. This implies that mentoring quality is positively correlated to mentor-mentee closeness (Aresi et al., 2021). The decision to choose a mentor or mentee rests with the two individuals in the programme. There must necessarily be a mentor and a willing mentee. This is a crucial factor for consideration in structured mentoring programmes to ensure mentor retention. In addition to quality relationships, observational mechanisms also enhanced mentoring success and professional practice accomplishments (Pryce et al., 2021).

Mentoring must ensure that the psychosocial needs and distinct circumstances of mentees are considered especially in the context of remote mentorship in a pandemic era (Pfund et al., 2021). A successful mentoring experience requires shared values, mutual respect, mutual trust, a sense of belonging and safety (Jefford et al., 2021). The mentor must necessarily have discipline competencies and empathy since most women deal with some of the challenges listed above.

For a mentoring programme to be successful, it must be structured. Regular meetings to assign tasks and evaluate progress must be in place. The mentee must be supported to apply for research and travel grants, awards, community engagement activities and conference attendance, among others.

Benefits of Mentoring for Women Scientists Across Sectors and Career Stages

One of the benefits of structured mentoring is professional development. Mentees' satisfaction, creation of an open environment for feedback and information dissemination, better understanding of the need for peer-support systems, research grants and other awards received and career progression are some of the benefits of structured mentoring programmes (Bredella et al., 2021). Quality mentoring has been linked to academic achievements and career outcomes. Mentees tend to mimic mentors in the professional space; behavioural attributes, strong work ethics and character are therefore a few of the many important attributes that the mentor must possess (Jablon & Lyons, 2021).

Mentoring and collaboration is rewarding for both the mentor and mentee. The sense of fulfilment that the mentor has seeing the mentee's career progression is priceless. Mentoring allows both the mentor and mentee to learn and grow together in the process. The mentee has professional guidance against career choice mistakes through the provision of psychosocial support. The mentee obtains the opportunity to be part of the mentor's research activities, networks and collaborations. The mentor also gets greater visibility from co-authored publications with the mentee in addition to professional recognition as an expert in that field of science. Connection to the mentor's networks, co-investigation of projects and co-authorship of articles, among others, provide enhanced career platform and visibility for the mentee. Mentoring is therefore desirable and beneficial for the mentor, mentee, professional communities and societies as summarized in Fig. 4.1.

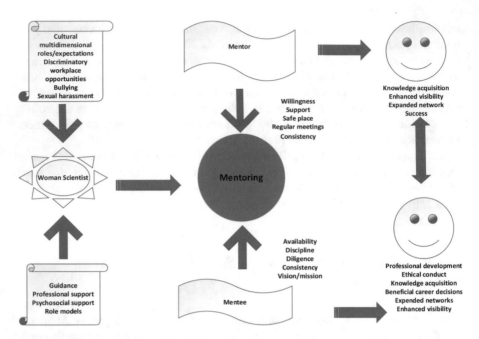

Fig. 4.1 Context, strategies and benefits of mentoring women scientists

4.12 Conclusion

The need for networking and collaboration in career progression and advancement of women scientists cannot be overemphasized. Mobility in terms of participation in local, regional and international meetings, conferences, workshops, symposia and many other career-related activities provides the platform for exchange of ideas and synergies for optimal performance. However, an inevitable event like the COVID-19 pandemic has disrupted these activities, making it pertinent for scientists to utilize the virtual space to their advantage for sustainable collaborations. In line with the 'new normal' operational globally, women scientists are encouraged to be more creative and innovative in their research designs so as to create enabling environments that will support sustainability and growth. In the same vein, scientific organizations need to include more career-enhancing opportunities with wider age limit to accommodate those whose career progression has been truncated due to family demands and raising of children. The role of established senior scientists in mentoring of early and mid-career women scientists is very crucial more especially with the global trend of advances in ICT to ensure that the 'leaky' STEMM pipeline is plugged in order to retain more women in the sciences. Finally, active membership in gender-specific and general scientific societies and very influential virtual presence in professional social media platforms will all be veritable tools to enhance inclusion and diversity for the promotion of women scientists in sustainable research and development.

References

Abramo, G., D'Angelo, A. C., & Murgia, G. (2017). The relationship among research productivity, research collaboration and their determinants. *Journal of Informetrics, 11,* 1016–1030.

Akyavuz, E. K., & Asıcı, E. (2021). The effect of volunteer management mentoring program on mentors' entrepreneurship tendency and leadership self-efficacy. *Participatory Educational Research, 8*(2), 1–16. https://doi.org/10.17275/per.21.26.8.2

Ansmann, L., Flickinger, T. E., Barello, S., Kunnemand, M., Mantwill, S., Quilligan, S., Zanini, C., & Aelbrecht, K. (2014). Career development for early career academics: Benefits of networking and the role of professional societies. *Patient Education and Counseling, 97,* 132–134.

ArauÂjo, E. B., Araujo, N. A. M., Moreira, A. A., Herrmann, H. J., & Andrade, J. S., Jr. (2017). Gender differences in scientific collaborations: Women are more egalitarian than men. https://doi.org/10.1371/journal.pone.0176791

Aresi, G., Pozzi, M., & Marta, E. (2021). Programme and school predictors of mentoring relationship quality and the role of mentors' satisfaction in volunteer retention. *Journal of Community and Applied Social Psychology, 31*(2), 171–183. https://doi.org/10.1002/casp.2495

Álvarez-Bornstein, M., & Bordons, M. (2021). Is funding related to higher research impact? Exploring its relationship and the mediating role of collaboration in several disciplines. *Journal of Informetrics, 15,* 101102.

Baltic gender https://www.baltic-gender.eu/outcomes;jsessionid=93D217F9FBBCBD75A4A9BDA029837F06. Best practices.

Bell, G. R., & Rosowsky, D. V. (2021). On the importance of mentorship and great mentors. *Structural Safety, 91.* https://doi.org/10.1016/j.strusafe.2021.102076

Bickel, J. (2007). The role of professional societies in career development in academic medicine. *Academic Psychiatry, 31,* 91–94.

Botham, C. M., et al. (2021). Biosciences Proposal Bootcamp: Structured peer and faculty feedback improves trainees' proposals and grantsmanship self-efficacy. *PLoS One, 15*(12 December). https://doi.org/10.1371/journal.pone.0243973

Bredella, M. A., et al. (2021). Radiology mentoring program for early career faculty— Implementation and outcomes. *Journal of the American College of Radiology, 18*(3), 451–456. https://doi.org/10.1016/j.jacr.2020.09.025

Butts, A., Wilber, J., & Rose, S. (2019). Help wanted!: A researcher's guide to utility-university collaborations. *The Electricity Journal, 32,* 106680. https://doi.org/10.1016/j.tej.2019.106680

Egbetokun, A., & Olofinyehun, A. (2021). Dataset on the production, dissemination and uptake of social science research in Nigeria. *Data in Brief, 35.* https://doi.org/10.1016/j.dib.2021.106932

Gaillard, J., Gaillard, A. M., & Arvanitis, R. (2014). Determining factors of international collaboration in science and technology: Results of a questionnaire survey. In G. Jacques & A. Rigas (Eds.), *Research collaboration between Europe and Latin America : Mapping and understanding partnership* (pp. 107–156). Ed. des Archives Contemporaines. ISBN 978-2-8130-0124-5.

Glăveanu, V. P., & de Saint, L. C. (2021). Social Media Responses to the Pandemic: What Makes a Coronavirus Meme Creative. *Frontiers in Psychology, 12,* 569987. https://doi.org/10.3389/fpsyg.2021.569987

Gazni, A., & Didegah, F. (2011). Investigating different types of research collaboration and citation impact: A casestudy of Harvard University's publications. *Scientometrics, 87*(2), 251–265.

Hansen, D. S. (2020). Identifying barriers to career progression for women in science: Is COVID-19 creating new challenges? *Trends in Parasitology, 36*(10), 799–801.

Iba, T., Levy, J. H., Connors, J. M., et al. (2020). The unique characteristics of COVID-19 coagulopathy. *Critical Care, 24,* 360. https://doi.org/10.1186/s13054-020-03077-0

International Science Council. (2021). Strengthening Science Systems. https://council.science/publications/strengthening-science-systems/

Jablon, E. M., & Lyons, M. D. (2021). Dyadic report of relationship quality in school-based mentoring: Effects on academic and behavioral outcomes. *Journal of Community Psychology, 49*(2), 533–546. https://doi.org/10.1002/jcop.22477

Jefford, E., et al. (2021). What matters, what is valued and what is important in mentorship through the Appreciative Inquiry process of co-created knowledge. *Nurse Education Today, 99*. https://doi.org/10.1016/j.nedt.2021.104791

Katz, S. J., & Martini, B. R. (1997). What is research collaboration? *Research Policy, 26*, 1–18.

Korbel, J. O., & Stegle, O. (2020). Effects of the COVID-19 pandemic on life scientists. Genome Biol, 21, 113. https://doi.org/10.1186/s13059-020-02031-1

Lee, S., & Bozeman, B. (2005). The impact of research collaboration on scientific productivity. *Social Studies of Science, 35*(5), 673–702.

Metcalfe, A. S., & Blanco, G. L. (2021). "Love is calling": Academic friendship and international research collaboration amid a global pandemic. *Emotion, Space and Society, 38*, 100763.

Moodley, K., & Gouws, A. (2020). *How women in academia are feeling the brunt of COVID-19. The Conversation Africa.* Funded by National Research Foundation (NRF) South Africa in collaboration with eight Universities in South Africa. www.theconversation.com. Accessed 14 Jan 2021.

Nüßer, L., Saunders, D., & Lynch, J. (2019). *Gender bias and equal opportunity in scientific research.* GLOBE, Session Summaries, 20(9). SETAC.

NAP. (1993). National Collaboratories: Applying Information Technology for Scientific Research. https://www.nap.edu/read/2109/chapter/2.

Opeolu, B., Arts, G. H. P., & Sabo-Attwood, T. (2018). SETAC Africa Women's Event – Calabar 2017. *Globe, 19*(2)., SETAC.

Pfund, C., et al. (2021). Reassess-Realign-Reimagine: A guide for mentors pivoting to remote research mentoring. *CBE Life Sciences Education, 20*(1), es2. https://doi.org/10.1187/cbe.20-07-0147

Pryce, J., et al. (2021). Understanding youth mentoring relationships: Advancing the field with direct observational methods. *Adolescent Research Review, 6*(1), 45–56. https://doi.org/10.1007/s40894-019-00131-z

Quadrio-Curzio, A., Blowers, T., & Thomson, J. (2020). Women, science and development: The leading role of OWSD. *Economia Politica, 37*, 1–12. https://doi.org/10.1007/s40888-020-00173-w

Radecki, J., & Schonfeld, R. C. (2020). *The impacts of COVID-19 on the research enterprise: A landscape review.* https://doi.org/10.18665/sr.314247

Richard, A. R., Liu, W., & Mukherjee, S. (2020). *The COVID-19 pandemic: A wake-up call for better cooperation at the science-policy-society interface*, Policy Brief, United Nations Department of Economic and Social Affairs, April 2020. https://www.un.org/development/desa/dpad/wpcontent/uploads/sites/45/publication/PB_62.pdf

Roehrl, R. A., Liu, W., & Mukherjee, S. (2020). UN/DESA Policy Brief #62: The COVID-19 pandemic: a wake-up call forbetter cooperation at the science–policy–society interface. https://www.un.org/development/desa/dpad/publication/un-desa-policy-brief-62-the-covid-19-pandemic-a-wake-upcall-for-better-cooperation-at-the-science-policy-society-interface/

Scopus. (2021). *Scopus® Expertly curated abstract & citation database.* Retrieved March 25, 2021, https://www.elsevier.com/solutions/scopus?dgcid=RN_AGCM_Sourced_300005030

SETAC. (2021). *SETAC Journals: Environmental quality through Science.* Available at setac.onlinelibrary.wiley.com. Assessed 28 Mar 2021.

Singh, R. P., Javaid, M., Kataria, R., Tyagi, M., Haleem, A., & Suman, R. (2020). Significant applications of virtual reality for COVID-19 pandemic. *Diabetes and Metabolic Syndrome: Clinical Research and Reviews, 14*(4), 661–664.

Smith, A. B., et al. (2021). Development and preliminary testing of the collaboration for leadership and innovation in mentoring survey: An instrument of nursing PhD mentorship quality. *Nurse Education Today, 98*. https://doi.org/10.1016/j.nedt.2021.104747

Sylverken, A. A., El-Duah, P., Owusu, M., et al. (2021). Transmission of SARS-CoV-2 in northern Ghana: Insights from whole-genome sequencing. *Archives of Virology*. https://doi.org/10.1007/s00705-021-04986-3

Xu, Y. J., & Martin, C. L. (2011). Gender differences in STEM disciplines: From the aspects of informal professional networking and faculty career development. *Gender Issues, 28*, 134. https://doi.org/10.1007/s12147-011-9104-5

Zhang, M., Zhang, G., Liu, Y., Zhai, X., & Han, X. (2020). Scientists' genders and international academic collaboration. An empirical study of Chinese Universities and Research Institutes. *Journal of Informetrics, 14*, 101068.

Chapter 5
Archiving Tool for Science and Scientist

Juliet Chinedu Alex-Nmecha and John Gibson Ogonu

5.1 Introduction

Science and scientists have impacted the world in different areas. Some are to alleviate aches and pains, help us to provide water for our basic needs – including our food, provide energy and make life more fun, including sports, music, entertainment and the latest communication technology, longer and healthier life, monitor our health, provide medicine to cure our diseases. Science generates solutions for everyday life and helps us to answer the great mysteries of the universe. In other words, science is one of the most important channels of knowledge. It has a specific role if when taken as an important one in the preservation of laboratory notes, important findings by women scientists, it means we will have vital documents to consult in future, as well as a variety of functions for the benefit of our society: creating new knowledge, improving education and increasing the quality of our lives.

Science equally responds to societal needs and global challenges. Public understanding, engagement with science and citizen participation including through the popularization of science are essential to equip citizens to make informed personal and professional choices. Governments need to make decisions based on quality scientific information on issues which necessitate the latest scientific knowledge such as health, agriculture, parliaments, etc. Governments need to understand the science behind major global challenges such as flood, climate change, ocean health, biodiversity loss and freshwater security. On the other hand, scientists must understand the problems policymakers face and endeavour to make the results of their research relevant and comprehensible to society. Scientists are important for the world because they help people understand the way the world works in very specific ways. Human beings have spent a lot of time figuring out how to stay alive and be happy, and science

J. C. Alex-Nmecha (✉) · J. G. Ogonu
Department of Library and Information Science, Faculty of Education, University of Port Harcourt, Port Harcourt, Nigeria
e-mail: juliet.alex-nmecha@uniport.edu.ng

© The Author(s), under exclusive license to Springer Nature Switzerland AG 2022
E. O. Nwaichi (ed.), *Science by Women*, Women in Engineering and Science, https://doi.org/10.1007/978-3-030-83032-8_5

has been a powerful tool for staying alive. Science and the scientist contribution to human advancement were made available through archives that collect and preserve original materials, software or knowledge, and thereby help scientists to get a glimpse into the past. Herein lies the added value of archival work: it captures and preserves what is not published and how scientific discoveries come about.

An archive is an accumulation of historical records – in any media – or the physical facility in which they are located. Archives contain primary source documents that have accumulated over the course of an individual or organization's lifetime and are kept to show the function of that person or organization (Galbraith, 1948). In general, archives consist of records that have been selected for permanent or long-term preservation on grounds of their enduring cultural, historical or evidentiary value. This means that archives are quite distinct from libraries with regard to their functions and organization, although archival collections can often be found within library buildings (*A Glossary of Archival and Records Terminology*, 2013). Therefore, if the science and scientist should be effective in their work, the role of an archivist in preserving scientific heritage, protecting authenticity, cataloguing, providing context, etc., which serves as useful tool and gateway to human development and advancement, should not be undermined.

This work considers the following: clarification, meaning, types of archive, the concept of science and scientist. The work shall also look at the role/tool of archives to science and scientists, general suggestions and conclusion.

5.2 Archive: The Meaning and Types

In general, archives consist of records that have been selected for permanent or long-term preservation on grounds of their enduring cultural, historical or evidentiary value. Archival records are normally unpublished and almost always unique, unlike books or magazines of which many identical copies may exist. An archive is a collection of old or historic records. (Glossary of Library and Internet Terms 2009). It is records that have been naturally and necessarily generated as a product of regular legal, commercial, administrative or social activities. An archive is an organized collection of the activities of a business, government, organization, institution or other corporate body, or the personal papers of one or more individuals, families or groups, retained permanently (or for a designated or intimidate periods of time) by their originator or successor for their permanent historical, informational or monetary value, usually in a repository managed and maintained by a trained archivist. Archive is defined as "the secretions of an organism" and is distinguished from documents that have been consciously written or created to communicate a particular message to posterity (Galbraith, 1948). According to a Latin dictionary, the English word archive is derived from the French archives, and in turn from the Latin archīum or archīvum, the Romanized form of the Greek ἀρχεῖον (arkheion) (2015). According to Murray, the Greek term originally referred to the home or dwelling of the Archon, a ruler or chief magistrate, in which important official state documents were filed and interpreted; from there its meaning broadened to encompass such concepts as "town hall" and "public records" (2009). The root of the Greek word is ἀρχή (arkhē), meaning, among other things, "magistracy, office, government", and derived from the verb ἄρχω (arkhō), meaning "to begin, rule, govern" (also the root of English words such as "anarchy" and "monarchy").

Archives were well developed by the ancient Chinese, the ancient Greeks and the ancient Romans (who called them Tabularia). However, they have been lost, since documents written on materials like papyrus and paper deteriorated at a faster pace, unlike their stone tablet counterparts. Archives of churches, kingdoms and cities from the Middle Ages survive and have often kept their official status uninterruptedly until now. They are the basic tool for historical research on these ages.

Modern archival thinking has many roots from the French Revolution. The French National Archives, who possess perhaps the largest archival collection in the world, with records going as far back as 625 AD, were created in 1790 during the Revolution from various government, religious and private archives seized by the revolutionaries (Archive: Definition, 2010).

5.3   Types of Archives

Walch (2006) identifies the following archives: government, academic, business (for-profit), non-profit and others.

Government Archive

Government archives include those maintained by local and state government as well as those maintained by the national (or federal) government. Anyone may use a government archive, and frequent users include reporters, genealogists, writers, historians, students and people seeking information on the history of their home or region. Many government archives are open to the public, and no appointment is required to visit (Directory of Corporate Archives, 2007). Some city or local governments may have repositories, but their organization and accessibility vary widely. In Nigeria, there are three government archives, namely National Archives of Nigeria situated at the three old regions of Ibadan, Enugu and Kaduna.

Academic Archive

These are archives in colleges, universities and other educational facilities that are typically housed within a library, and duties may be carried out by an archivist. Maher states that academic archives exist to preserve institutional history and serve the academic community (1992). An academic archive may contain materials such as the institution's administrative records, personal and professional papers of former professors and presidents, memorabilia related to school organizations and activities, and items the academic library wishes to remain in a closed-stack setting, such as rare books or thesis copies. Access to the collections in these archives is usually by prior appointment only; some have posted hours for making inquiries. Users of academic archives can be undergraduates, graduate students, faculty and staff, scholarly researchers and the general public.

Business (For-Profit) Archive

These archives located in for-profit institutions are usually those owned by a private business. Examples of business archives include Coca-Cola (which also owns the separate museum World of Coca-Cola), Procter & Gamble, Motorola Heritage Services and Archives and Levi Strauss & Co. These corporate archives maintain historic documents and items related to the history and administration of their companies. Business archives serve the purpose of helping their corporations maintain control over their brand by retaining memories of the company's past. Especially in business archives, records management is separate from the historic aspect of archives. These archives are typically not open to the public and only used by workers of the owner company, though some allow approved visitors by appointment. Business archives are concerned with maintaining the integrity of their company and are therefore selective of how their materials may be used (Michelle, 2005).

Church

Church archives such as the Vatican Secret Archive, archdioceses, dioceses and parishes also have archives in the Roman Catholic and Anglican Churches. Whitehill (1962) opines that very important are monastery archives, because of their antiquity, like the ones of Monte Cassino, Saint Gall and Fulda. The records in these archives include manuscripts, papal records, local Church records, photographs, oral histories, audiovisual materials and architectural drawings. Most Protestant denominations have archives as well, including the Presbyterian Historical Society, the Moravian Church Archives, the Southern Baptist Historical Library and Archives, the United Methodist Archives and History Center of the United Methodist Church and the Christian Church (Disciples of Christ).

Non-profit Archive

Non-profit archives include those in historical societies, not-for-profit businesses such as hospitals and the repositories within foundations. Non-profit archives are typically set up with private funds from donors to preserve the papers and history of specific persons or places. These institutions rely on grant funding from the government as well as the private funds. Depending on the funds available, non-profit archives may be as small as the historical society in a rural town to as big as a state historical society that rivals a government archive. Creigh and Pizer (1991) point that users of this type of archive may vary as much as the institutions that hold them. Employees of non-profit archives may be professional archivists, para-professionals or volunteers, as the education required for a position at a non-profit archive varies with the demands of the collection's user base.

Web Archive

Web archiving is the process of collecting portions of the World Wide Web and ensuring the collection is preserved in an archive, such as an archive site, for future researchers, historians and the public. Due to the massive size of the Web, web archivists typically employ web crawlers for automated collection. Similarly, software code and documentation can be archived on the Web which in turn serves as a veritable for scientists

Individual/Personal Archive

This is a special type of archive built by individuals in order to preserve, capture and guard their personal papers and other documentary output of an individual concerned for future use. Organizing personal collections can be a way to tell a story about all one did in life as regards findings got in previous years' experiments that yielded beautiful results. When a woman scientist keeps such records, another female scientist will benefit from it when the need arises.

According to Kells in Ashefelder (2016),

"Archive is good
   What you save in the archive gives you a sense of people
   And the person who saved it, even if there's not much
   To the document kept, it shows you what they valued,
   What they worked on and all that transpired for a project
   to be successful"

5.4   The Concept of Science

Science is the systematic study of the nature and behaviour of the material and physical universe, based on observation, experiment and measurement, and the formulation of laws to describe these facts in general terms. According to Harper (2016), science (from the Latin word scientia, meaning "knowledge") is a systematic enterprise that builds and organizes knowledge in the form of testable explanations and predictions about the universe (Wilson, 1999). Science is based on research, which is commonly conducted by scientists working in academic and research institutions, government agencies and companies. The practical impact of scientific research has led to the emergence of science policies that seek to influence the scientific enterprise by prioritizing the development of commercial products, armaments, health care, public infrastructure and environmental protection.

Modern science is typically divided into three major branches that consist of the natural sciences (e.g. biology, chemistry and physics), which study nature in the broadest sense; the social sciences (e.g. economics, psychology and sociology), which study individuals and societies; and the formal sciences (e.g. logic, mathematics and theoretical computer science), which study abstract concepts. There is disagreement, according to Fetzer (2013), however, on whether the formal sciences actually constitute a science as they do not rely on empirical evidence. Disciplines that use existing scientific knowledge for practical purposes, such as engineering and medicine, are described as applied sciences (Fischer & Fabry, 2014).

Science is based on research, which is commonly conducted by scientists working in academic and research institutions, government agencies and companies, This research has led to the emergence of science policies that seek to influence the scientific enterprise by prioritizing the development of commercial products, armaments, health care, public infrastructure and environmental protection.

5.5   Branches of Science

Modern science is commonly divided into three major branches: natural science, social science and formal science. Bunge (1998) states that each of these branches comprises various specialized yet overlapping scientific disciplines that often possess their own nomenclature and expertise. Both natural and social sciences are empirical sciences, as their knowledge is based on empirical observations and is capable of being tested for its validity by other researchers working under the same conditions (Popper, 2002). There are also closely related disciplines (Table 5.1) that use science, such as engineering and medicine, which are sometimes described as applied sciences.

Table 5.1 Relationships between the branches of science (Bunge, 1998)

	Science		
	Empirical sciences		
	Natural science	Social science	Formal science
Basic	Physics, chemistry, biology,science and space science	Anthropology, economics, political science, sociology, human geography and psychology	Logic, mathematics and statistics
Applied	Engineering, agricultural science, medicine and materials science	Business administration, public policy, marketing, law, pedagogy and international development	Computer science

5.6   Impacts of Science

Discoveries in fundamental science can be world changing. Popper (2002) tabulated the impact science has on society as shown in Table 5.2.

5.7   Scientists

A scientist is a person who conducts scientific research to advance knowledge in an area of interest. In classical antiquity, there was no real ancient analogue of a modern scientist. Cyranoski et al. (2011) state that philosophers engaged in the philosophical study of nature called natural philosophy, a precursor of natural science. Kwok (2017) opines that the term scientist was coined by William Whewell in 1833. In modern times, many professional scientists are trained in an academic setting and, upon completion, attain academic degree, with the highest degree being a doctorate such as a Doctor of Philosophy (PhD) (Woolston, 2007). Many scientists pursue careers in various sectors of the economy such as academia, industry, government and non-profit organizations (Lee et al., 2007). Scientists exhibit a strong curiosity about reality, with some scientists having a desire to apply scientific knowledge for the benefit of health, nations, environment or industries. Other motivations include recognition by their peers and prestige.

5.8   Archive Roles/Tool for Science and Scientists

The organized collection of noncurrent records of activities of a business, government, personal papers of one or more individuals, groups especially scientists on inventions, discoveries or laboratory experiments serves as bedrock for future invention and advancement in knowledge. These roles of archivists are useful tool for science and scientists and can be further summed under the followings.

Table 5.2 Impact of science on the society

Research	Impact
Static electricity and magnetism (c. 1600) Electric current (18th century)	All electric appliances, dynamos, electric power stations, modern electronics, including electric lighting, television, electric heating, transcranial magnetic stimulation, deep brain stimulation, magnetic tape, loudspeaker and the compass and lightning rod
Diffraction (1665)	Optics, hence fibre-optic cable (1840s), modern intercontinental communications and cable TV and Internet
Germ theory (1700)	Hygiene, leading to decreased transmission of infectious diseases; antibodies, leading to techniques for disease diagnosis and targeted anticancer therapies
Vaccination (1798)	Leading to the elimination of most infectious diseases from developed countries and the worldwide eradication of smallpox
Photovoltaic effect (1839)	Solar cells (1883), hence solar power, solar-powered watches, calculators and other devices
The strange orbit of Mercury (1859) and other researchleading to special (1905) and general relativity (1916)	Satellite-based technology such as GPS (1973), satnav and satellite communications
Radio waves (1887)	Radio had become used in innumerable ways beyond its better-known areas of telephony and broadcast television (1927) and radio (1906) entertainment; other uses included – emergency services, radar (navigation and weather prediction), medicine, astronomy, wireless communications, geophysics and networking; radio waves also led researchers to adjacent frequencies such as microwaves, used worldwide for heating and cooking food
Radioactivity (1896) and antimatter (1932)	Cancer treatment (1896), radiometric dating (1905), nuclear reactors (1942) and weapons (1945), mineral exploration, PET scans (1961) and medical research (via isotopic labelling)
X-rays (1896)	Medical imaging, including computed tomography
Crystallography and quantum mechanics (1900)	Semiconductor devices (1906), hence modern computing and telecommunications including the integration with wireless devices: the mobile phone,[1] LED lamps and lasers
Plastics (1907)	Starting with Bakelite, many types of artificial polymers for numerous applications in industry and daily life
Antibiotics (1880s, 1928)	Salvarsan, penicillin, doxycycline, etc.
Nuclear magnetic resonance (1930s)	Nuclear magnetic resonance spectroscopy (1946), magnetic resonance imaging (1971), functional magnetic resonance imaging (1990s)

5.9   Protecting Authenticity

The authenticity of original documents is central to archives: once authenticity can no longer be proven, it is very difficult to re-ascertain it which may lead to distortion and baseless precedence in the future. Niu (2016) stated that preserving authenticity requires several factors. First, ensuring that an archived document, regardless of its format – digital or analogue – is indeed the one that was deposited. Archived documents, such as administrative papers, laboratory books or datasets, need to be indubitably what they claim to be at any point in time they are consulted. Through protecting authenticity, archived documents are able to prove that a past action has indeed taken place, which can be crucial in research. In current scientific archives, protecting authenticity is also relevant for protecting intellectual property and managing scientific information. For example, being able to demonstrate that a certain experiment was conducted during a certain timeframe and making sure that it is put in context within a larger set of materials is important for demonstrating precedence or ownership of an idea. Similarly, current discussions about data management also revolve around the notion of preserving authenticity and providing trustworthy datasets through their demonstrable authenticity. Archivists can therefore contribute to the management of scientific data to ensure that material is available in the long term and is trustworthy. Certainly, having noted the similarities of goals, archivists and data curators might be able to "collaborate, share expertise in knowledge organization and integrate resources" in order to protect the authenticity of scientific material which will serve as a point of reference in further scientific discovery and invention by the consulting scientists.

5.10   Providing Context

It is worthy of note that the role of archivists in providing contest is a major one, the presentation of timeline archival material and historical publications from a number of institutions and provides access to contextualized, first-hand documents. (Such resources ensure that archived material is put in the context of collections, institutions and topics, all of which are important to protect and communicate archival integrity. Whether this is done at the level of catalogues or through national and international initiatives, the objective is always to create and control relationships between materials while enabling researchers to effectively navigate through resources in order to find what they are looking for. As such, the archivist's task of providing context to collections is one of collaboration, thinking beyond single subject areas and enriching archival finding aids. As scientists collaborate, so do archivists with each other and with scientists to provide adequate contextual information within and beyond their own institutions which enhances effective scientific research and development.

5.11   Preserving the Scientific Heritage

The International Human Genome Sequencing Consortium included some 20 organizations, and managing this massive scientific collaboration generated a lot of communication, which was mostly conducted electronically. According to Ferry, these emails allow us to see into the realities of a large-scale scientific partnership, with all its ups and downs (2013). For example, in October 1998, one email had the subject line "Friendly fire". Written by John Sulston, then director of the Sanger Institute, to Francis Collins, who was, at the time, working at the US National Human Genome Research Institute, this emotionally charged email is part of a broader conversation about communication and suggested changes within the project, partly in response to the launch of Celera by Craig Venter earlier that year to sequence the human genome as a commercial venture. Sulston's email bears witness to how science is also about managing human relationships, which is not obvious in peer-reviewed publications. Through its work, the HGAP ensures that such evidence of daily interactions, personalities, ideas and false starts behind the human genome project are not lost forever (Ferry, 2013).

Capturing this kind of material is not new, though, even if the medium of emails is new, electronic mailing lists and instant messaging are today's equivalent of the Enlightenment's Republic of Letters, field notebooks and learned societies of the past. Since science has always thrived on exchange of ideas, these communications shed light on the mechanisms of science at any time. By archiving them, scientists not only record their results but how these were obtained and what was needed to achieve success. Archives therefore help scientists, historians and other scholars make this material available for present and future research by collecting, preserving and cataloguing physical and digital evidence. Emails, electronic mailing lists and instant messaging are today's equivalent of the Enlightenment's Republic of Letters, field notebooks and learned societies of the past. Whether at research institutes or universities, the number of archivists employed to preserve scientific heritage is growing. Archivists are essential for creating such a resource. At the core of their tasks is the preservation of material in an "unaltered and unmediated and unbroken context" and protecting its continuous custody (Cook, 2013). They provide context through cataloguing, which adds value to the material and makes it easier to find particular documents. In addition, the preservation of analogue and digital material ensures that material is usable by future researchers and generations.

5.12   Cataloguing

Cataloguing is perhaps the most obvious task, regardless of the nature of the material itself. The creation of a catalogue is not an end in itself, but part of intellectual control, which involves tools such as finding aids, indexes or other guides that help researchers to identify and locate pertinent archived material. Cataloguing is a multilayered activity that needs to consider the past, the present and the future to develop a robust resource that will facilitate easy access to material over time. The past is relevant because of the principle of original order, discussed above: it preserves the context of records by showing how they were ordered and used by their creator. Cataloguing also considers the present because it reflects the structures or settings within which a catalogue is created through the application of certain standards or the use of terminology. Finally, it must consider the future, since future users will need to be able to access material efficiently, through a resource adapted to their time. In this regard, one evident challenge of cataloguing scientific archives is the fact that science moves quickly and so do the terms that it uses.

To ensure that material remains findable even when subject matters have been re-named or taxonomies altered requires ongoing work and sometimes starting all over again. This is not unlike the work of classifying extinct and extant organisms through their biological relationships to one another, a continuous challenge which is taken on by taxonomy. As science advances and methods change, if organisms are reclassified, access to the earlier literature and documents nevertheless needs to be maintained. Women being natural managers and by virtue of their scientific stance will do better in safeguarding their information materials, like their lab notes with all that they have written down to help for future use of more researches; moreover, if the search is confusing, the materials preserved in the archives will do the wonders of clarifications. So too with archives: archivists need to ensure continued access to and findability of material through indexing, creating subject guides and, most importantly, communicating with the producers and users of material regularly to receive feedback on the catalogue.

5.13 Preservation

Archivists have developed and improved methods to preserve and conserve the physical integrity of analogue material across time, sometimes going to extraordinary lengths to protect it against environmental or political threats. Preventive measures and care help to ensure that material from papers, wax seals, comic books, photographs or parchment does not degrade faster than it can be cared for (Brenner & Roberts, 2007).

Today's notes, patents and manuscripts are "recorded in ephemeral electronic media that are far too easily lost with the push of a button or the failure of a hard drive". If these are not coherently preserved, they will be lost forever. The term refers to the idea that if digital material is not actively preserved, no traces will be available for future research on the extremely productive and innovative decades. This problem is not only limited to text files but also photographs, video and audio media. No institution or person is immune to the loss of digital material: infamous. According to Blakeslee (1990), major institutions in Europe lost their important data from several space programmes because tapes were poorly stored. They went to great lengths and incurred considerable expenses to recover data and photographs of our solar system. Furthermore, being able to access digital material is not simply about maintaining hardware but also functioning software, including licences and passwords if necessary. As scientists collaborate, so do archivists with each other and with scientists to provide adequate contextual information within and beyond their own institutions.

The field of digital preservation is rife with recent developments to prevent the loss of data and ensure long-term and trustworthy access to digital material. These include migration to newer systems and servers, replication and emulation, all of which have their benefits and shortfalls. Code and software can be archived too and shared through such initiatives as Software Heritage to scientist to further their usefulness in discovery and invention.

5.14 Transfer of Knowledge/Bridging the Gap

Archives aid the researchers to transfer science, technology and innovation knowledge and its application from one generation to another, thereby driving our pursuit of more equitable and sustainable development. Again, the laboratory has recognized the value of their knowledge by sometimes trying to keep the same administrative assistants in areas experiencing leadership changes, so the new leaders can draw on their experience. For the lab's archivists and others interested in the history of the lab's operations, these administrative assistants are repositories of unique institutional knowledge that scientific and technical staff may not know. However, this kind of valuable information will often not be found in the records produced by support staff. The best way to capture it is usually through oral interviews, which make their memories and observations part of the archival record.

Subsequently, the importance of scientific archives for molecular biology in particular was brought up a decade ago by Sydney Brenner and Richard J. Roberts in a letter to *Nature* (Newton, 1990). Newton the great scientist (Newton, 1990) puts it, "If I have seen further, it is by standing on the shoulders of giants", referring to how his own discoveries had built on the work of other scientists like René Descartes and Robert Hooke. Addressing their fellow molecular biologists and bioscientists, they argued: "Let's not wait until memories have faded and papers been discarded at the end of a career before deciding to save our heritage". The legacy in danger of being lost is not the published record, which is preserved and made accessible through libraries and publishers, but the material that complements it: laboratory notebooks, email exchanges or prototype instruments. This is the evidence of how science is actually being done – the closest to conversations in the laboratory that led to a new invention or a groundbreaking discovery. Herein lies the added value of archival work: it captures and preserves what is not published and how scientific discoveries come about.

5.15 Conclusion

The archives of science document are one of humankind's greatest achievements. The scientific method seeks to expand on the discoveries of the past, with each generation of scientists building upon the work of their predecessor, or as Newton the great scientist highlighted the importance of leaning on past efforts to push boundaries, women in science (and scientists generally) need to embrace the use of archives to preserve their records of experimental findings. By working together, archivists and scientists can help ensure that today's traces remain available for others to consult and build on. Archivists ensure continued access to today's science tomorrow, regardless of format. Complementing the published record, the archives of science provide unique insights into the sequence of events between individuals, across institutions and collaborations. Today's science is based on the archival provisions and tool of yesterday scientists retrieved and consulted from.

5.16   General Suggestions

Government should set up more archival centre in the nation.

The employee should be trained to function.

The archival centres should be equipped with modern facilities to enable scientists and researchers satisfy their needs.

Collaboration between archivists, scientists and other stakeholders should be encouraged to advance the frontier of knowledge.

Women scientists, individuals and corporate organizations should be encouraged to venture into archival business.

Collaboration among archivists, librarians, historians and scientists should be fostered to sustain to tomorrow's knowledge advancement and inventions.

References

Archīum. Archived 24 September 2015 at the Wayback Machine, Charlton T. Lewis, Charles Short, A Latin Dictionary, on Perseus

Archive: Definition, Synonyms from. Answers.com. Archived from the original on 23 May 2010. Retrieved 1 June 2010.

Archived from the original on 22 October 2013. Retrieved 21 October 2013

Ashenfelder, M. (2016). *Your personal archiving project: Where do you start?* Retrieved from library of congress>blog>signal>your personal archiving project: where do you start.

Blakeslee, S. (1990). Lost on earth: Wealth of data found in space. *The New York Times*, 20 March, C13.

Brenner, S., & Roberts, R. J. (2007). Save your notes, drafts and printouts: Today's work is tomorrow's history. *Nature, 446*, 725. [PubMed] [Google Scholar]

Bunge, M. A. (1998). *Philosophy of science: From problem to theory* (p. 24). Transaction Publishers. isbn:978-0-7658-0413-6.

Cook, T. (2013). Evidence, memory, identity, and community: Four shifting archival paradigms. *Archival Science, 13*, 106. [Google Scholar]

Creigh, D. W., & Pizer, L. R. (1991). *A primer for local historical societies* (2nd ed., p. 122). American Association for State and Local History. ISBN 9780942063127.

Cyranoski, D., Gilbert, N., Ledford, H., Nayar, A., & Yahia, M. (2011). Education: The PhD factory. *Nature, 472*(7343), 276–279.

Directory of Corporate Archives. Hunter information.com. Archived from the original on 5 April 2007. Retrieved 8 May 2007.

Ferry, G. (2013). Science today, history tomorrow. *Nature, 493*, 19. [PubMed] [Google Scholar]

Fetzer, J. H. (2013). Computer reliability and public policy: Limits of knowledge of computer-based systems. In *Computers and cognition: Why minds are not machines* (1st ed., pp. 271–308. ISBN 978-1-443-81946-6). Kluwer Academic Publishers.

Fischer, M. R., & Fabry, G. (2014). Thinking and acting scientifically: Indispensable basis of medical education. *GMS Zeitschrift für Medizinische Ausbildung, 31*(2). https://doi.org/10.3205/zma000916. PMC 4027809.

Galbraith, V. H. (1948). *Studies in the public records* (p. 3). T. Nelson.

Glossary of Archival and Records Terminology. Society of American Archivists.

Harper, D. (2016). *Science*. Online Etymology Dictionary. Retrieved September 20, 2014.

Kwok, R. (2017). Flexible working: Science in the gig economy. *Nature, 550*, 419. https://doi.org/10.1038/nj7677-549a

Lee, A., Dennis, C., & Campbell, P. (2007). Graduate survey: A love–hurt relationship. *Nature, 550*(7677), 549–552. https://doi.org/10.1038/nj7677-549a

Maher, W. J. (1992). *The Management of College and University Archives*. Society of American Archivists and the Scarecrow Press. OCLC 25630256.

Michelle, R. (2005). The correlation of archival education and job requirements since the advent of encoded archival description. *Journal of Archival Organization, 3*(1), 61–79. (Accessed 23 July 2014).

Murray, S. (2009). *The Library: An illustrated history* (p. 7). Skyhorse Publishing. ISBN 978-1-61608-453-0.

Newton, I.: Letter to Robert Hooke. 5 February 1675. Simon Gratz autograph collection [0250A]. Historical

Niu, J. (2016). Aggregate control of scientific data. *Archives and Records, 37*, 62. [Google Scholar]

Popper, K. R. (2002). *A survey of some fundamental problems. The Logic of Scientific Discovery* (pp. 3–26). Routledge Classics. ISBN 978-0-415-27844-7. OCLC 59377149.

Sulston, J., Mallet, F., Staden, R., Durbin, R., Horsnell, T., & Coulson, A. (1988). Software for genome mapping by fingerprinting techniques. *Bioinformatics, 4*, 125–132. [PubMed] [Google Scholar]

Walch, V. I. (2006). "Archival census and education needs survey in the United States: Part 1: Introduction" (PDF). *The American Archivist, 69*(2), 294–309. Archived (PDF) from the original on 14 March 2007. Retrieved 30 April 2007.

Whitehill, W. M. (1962). Introduction. In *Independent historical societies: an enquiry into their research and publication functions and their financial future* (p. 311). Boston, Massachusetts.

Wilson, E. O. (1999). "The natural sciences". Consilience: The unity of knowledge (Reprint ed.). : Vintage. pp. 49–71. ISBN 978-0-679-76867-8.

Woolston, C. (2007). Editorial (ed.). "Many junior scientists need to take a hard look at their job prospects". *Nature, 550*, 549–552. https://doi.org/10.1038/nj7677-549a

Chapter 6
Wearing Our Gender Lens in Research Design and Development

Blessing Adanta Odogwu

6.1 Introduction

The phrase 'wearing gender lens in research' is an informal expression that connotes integrating or mainstreaming a gender perspective during the development, implementation and evaluation of a research. This means that the research should have a gender-responsive content, by not only identifying gender issues (or biases) but also proffering solutions to them. To learn how to properly wear a gender lens or mainstream gender in research, it will be better to first understand what gender, gender responsive and gender mainstreaming in research mean (AWARD, 2014).

6.2 What Is Gender?

The word 'gender' has sometimes been erroneously attributed to *only* women, but in the real sense, it has to do with the relationships and roles played between and among men and women (Udry, 1994; AWARD, 2014). A commonly accepted definition for 'gender' refers to 'the roles, behaviors, activities and attributes that a given society at a given time considers apt for men and women' (AWARD, 2014). According to UNESCO Gender Lens (2003), the gender designations were socially created and their roles are still evolving!) and generated based on the beliefs and social constructs about manhood and womanhood. These gave rise to the men and women roles, responsibilities, activities, access to and control over

B. A. Odogwu (✉)
Department of Plant Science and Biotechnology, University of Port Harcourt,
Port Harcourt, Nigeria
e-mail: Blessing.odogwu@uniport.edu.ng

© The Author(s), under exclusive license to Springer Nature
Switzerland AG 2022
E. O. Nwaichi (ed.), *Science by Women*, Women in Engineering and Science,
https://doi.org/10.1007/978-3-030-83032-8_6

resources and decision-making opportunities that have established resilient norms regarding what is expected, allowed and valued for women, men, girls and boys, thus shaping their socialization and institutions such as the family, media, law and education system; how race, class, age, religion, disability and sexuality are lived; and the ways in which inequality is experienced. Therefore, both women and men 'experience relations of gender from radically different positions of personal, social, economic, and political power' – this often results in discrimination (UNESCO Gender Lens, 2003).

A history of discrimination and restraining roles is unconsciously written into everyday routines and policies. Gender is not only a socially constructed definition of women and men but also a socially constructed definition of the relationship between the sexes. This construction contains an unequal power relationship with male domination and female subordination in most spheres of life. Men and the tasks, roles, functions and values contributed to them are valued – in many aspects – higher than women and what is associated with them. It is increasingly recognized that the society is characterized by this male bias: that is, the male norm is taken as the norm for society as a whole, which is reflected in policies and structures that often unintentionally reproduce gender inequality.

Therefore, to distribute resources and responsibilities between men and women, it is pertinent to take into consideration the gender roles and relationships of a community. More so, to integrate gender in research and development, there is a need to look beyond the observed roles played by the different gender groups, but to explore the societal systems, structures and power relations that enforce them or form a barrier that may prevent a gender group from accessing the research outputs. For this reason, there is a need to understand what a gender-responsive research will entail.

6.3   Gender-Responsive Research

Gender-responsive research is a study that addresses gender issues such as the differences in the conditions, needs, participation, access to resources and development, control of assets and decision-making powers between the assigned roles of males and females (AWARD, 2014). It explores the significance of the roles assigned to males and females in all areas and covers the analysis of priorities and potential outcomes of a given planned research or development project, subjects or samples selected for a given study and the gender dynamics in the participating institution(s) and among the project or study teams. This implies that for a research to be gender responsive, the researcher should be knowledgeable about gender, has identified key gender interest groups and the relevant gender issues both in the research questions and in the team implementing the research and is familiar with the tools used to collect and analyse gender-disaggregated data. That means the researcher should wear a gender lens.

6.4   What Is Gender Mainstreaming?

The United Nations Economic and Social Council, defines gender mainstreaming as 'the process of assessing the implications for women and men of any planned action, including legislation, policies or programmes, in all areas and at all levels'. It further stated that 'it is a strategy for making women's as well as men's concerns and experiences an integral dimension of the design, implementation, monitoring and evaluation of policies and programmes in all political, economic and societal spheres so that women and men benefit equally, and inequality is not perpetuated, with the ultimate goal of achieving gender equality' (Hosein et al., 2020).

The term 'gender mainstreaming' was first used in 1985 at the Third World Conference on Women, which took place in Nairobi. It was later adopted in 1995 as a strategy to include a gender perspective in legislation, policies, programmes and projects by the Platform for Action at the Fourth World Conference on Women, which took place in Beijing. Later in 1997, the United Nations adopted the first resolution on gender mainstreaming to guide and support Member States in the implementation of global commitments related to gender equality and the empowerment of women. On that occasion, Member States agreed to assess the differentiated implications, for women and men, of any planned action, including legislation, throughout the entire cycle of policies and programmes from the design phase to the evaluation process (Hosein et al., 2020).

More recently, gender mainstreaming has gained a new stimulus with the adoption in 2015 of the 2030 Agenda for Sustainable Development and the 17 Sustainable Development Goals (SDGs). The new agenda emphasized the importance of systematic mainstreaming of a gender perspective in its comprehensive implementation, since the realization of gender equality and the empowerment of women and girls will make a decisive and cross-cutting contribution to progress in all the goals and targets, in particular in this remaining decade to successfully implement this global agenda (Hosein et al., 2020; Gender mainstreaming, 1998; ECLAC, 2020).

Therefore, gender mainstreaming starts with the recognition that gender equality is a basic development goal and a key objective of research and development planning. It is therefore a strategy that ensures that there is an enabling environment for both women and men to have access to and control over resources, decision-making and benefits at all stages of the development process in ways that promote human rights, gender equality and gender equity.

6.5   Why Is It Important to Wear a Gender Lens When Designing and Developing Our Research?

As mentioned earlier, gender-responsive research looks beyond the observed roles played by men and women, but explores the societal systems, structures and power relations that enforce them, that is, the societal values and norms of a given time in the history of a community (AWARD, 2014). The importance of integrating gender into research planning, implementation and evaluation has been highlighted by Odogwu (2020). According to Odogwu (2020), the four reasons why a researcher should wear a gender lens when designing and developing their research are as follows:

(i) **Research** design improvement and focus: Putting on a gender lens or developing a gender-responsive research will enable the researchers to identify potential stakeholders and beneficiaries of their research value chain, identify their needs and research gaps and determine the possible research outputs that will meet these specific needs, thereby streamlining the research focus.

(ii) **Ease of** technology transfer **from lab to the field**: Wearing the gender lens will make it easy for the researchers to predict the impact of their research outcomes, engender ease in adoption of technology and even ascertain the gender dynamics and barriers that can affect the uptake of the technology (this will be important for re-strategizing the research focus if the need arises) and where possible enhance the adaptive capability of the gender groups that less likely benefit from the technology but do not have access to it.

(iii) Network improvement: When designing and developing a research, donning the gender lens will help the researcher identify potential collaborators that will enable him or her to expand their networks. These networks are invaluable during and after the research project.

(iv) **Funders focus and access to funding for research**: Most funders and donors have target gender groups or SDGs linked to gender group(s) of interest! Designing a research project with a gender perspective will make it easier for a researcher to access these funds.

6.6 How to Properly Wear the Gender Lens?

During the research conceptual or planning stage, a researcher can integrate a gender perspective by conducting a gender analysis. After identifying the gender groups to be most impacted by the research, the researcher can engage the identified gender groups from the research planning stage to the evaluation stage. The involvement and participation of the gender groups at the beginning of the research process will make it easy for them to adopt the research outputs or technology (Assefa & Roo, 2015; Leong et al., n.d.).

6.7 Gender Analysis

Gender analysis is a study that examines the different roles and responsibilities of women and men and how these affect society, culture, the economy and even politics. For example, important differences exist between women and men in their quality of life; in the amount, kind and recognition of work they do; in health and literacy levels; and in their economic, political and social standing. Women are too often marginalized in their families and their communities, suffering from a lack of ⁏⁏⁏⁏⁏ ⁏⁏ ⁏⁏⁏⁏⁏⁏ ⁏⁏⁏⁏, ⁏⁏⁏⁏⁏⁏⁏⁏⁏⁏ ⁏⁏⁏⁏⁏⁏⁏⁏ ⁏⁏⁏⁏⁏⁏⁏ ⁏⁏⁏⁏⁏ ⁏⁏⁏ ⁏⁏⁏⁏⁏⁏ ⁏⁏ ⁏⁏⁏⁏⁏ According to Assefa and Roo (2015) and Leong et al. (n.d.), the main goals of gender analysis are:

(a) To create a 'gender looking-glass' through which we examine and better understand the communities.
(b) To provide evidence to make decisions and implement the project/programme that promotes gender equity.
(c) To better understand the opportunities/problems in the community and plan interventions which are beneficial to both women and men.
(d) To find the best strategies and solutions to address the different needs and dynamics of men and women living in poverty.

6.8 When Should Gender Analysis Be Used?

Gender analysis can be conducted any time a researcher is looking for ways to better understand and improve a community of interest. The best opportunities to do gender analysis are during the initial design of a project, before the implementation of a policy or during the evaluation of a project or policy (Assefa & Roo, 2015). To conduct an effective gender analysis, both traditional and non-traditional research methods can be used to collect data. The traditional method for data collection includes formal interviews and surveys, mapping and research through libraries and organizations, while the non-traditional method includes household interviews and focus group sessions, informal conversations, walking tours observing community practices and other methods where there is participation by a diverse group of people (Leong et al., n.d.). To make the process of collecting data easier and smarter, the use of gender analysis tools is recommended.

6.9 Gender Analysis Tools

There are four types of gender analysis tools that researchers can employ. They are:

(i) **Gender** situational analysis **(or Harvard model) framework:** It is one of the most commonly used gender analysis frameworks. It is developed based on the understanding that women and men are affected by development activities differently. This is a tool used for establishing the gender categories, relations and issues of a given community at a certain time and location. This framework helps researchers identify specific gender groups or issues, that is, identifying specific problems or opportunities for a specific gender. The tool consists of

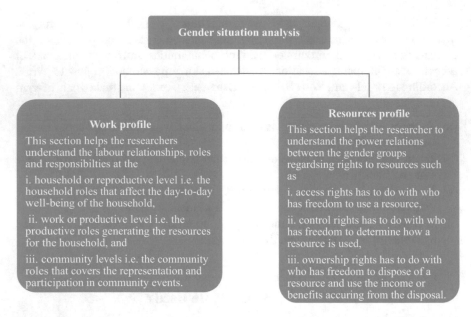

Fig. 6.1 The two sections considered in the gender situational analysis framework (AWARD, 2014)

two distinct sections, namely the work profile and resources profile that influence events (Fig. 6.1). The gender situation analysis often uncovers the challenges and opportunities that the researcher may wish to explore further in the research (AWARD, 2014; Assefa & Roo, 2015; Leong et al., n.d.).

(ii) Problem and opportunity analysis: This analysis is usually a follow-up of the situational analysis. Among the different tools used in the analysis (Fig. 6.1), the most common tool is the preference ranking. This tool is based on the different issues identified that affect the different gender groups such as their perceptions or choices, problems and opportunities the gender groups would make or face independently. The choices of different gender groups are based on their roles and responsibilities (AWARD, 2014).

(iii) **Gender analysis matrix (GAM)**: The gender analysis matrix (GAM) is an analytical tool that is used to determine the differentiated impact that an intervention might have on women and men (Assefa & Roo, 2015). It assesses the impact with respect to the positive and negative changes that a given intervention has brought in terms of labour, resources, time and sociocultural factors. The tool comprehends the impact in terms of labour (whether the new technology is more (or less) labour demanding), resource (whether the new technology is more (or less) resource intensive as compared to the conventional), time (whether the new technology is time taking or time saving) and cultural perspective (the changes in social aspects of the people's life as a result of the intervention).

Culture: has to do with underlying values, practices and norms and their implications for options prioritized by the gender groups.

Equity: has to do with access to and use of benefits, that is, which gender group will actually benefit from the research output(s).

Technical competencies: has to with any special skills required and the implications for participation and access to benefits by the gender groups.

Time to benefit: Has to do with how long it will take to realize benefits and how this affects each gender group.

Costs: has to do with how much the project will cost the community and the ability of each gender group to pay (low-cost options).

Location: has to do with how the location or venue of the project will affect access, especially by those with limited freedom of mobility.

Magnitude of benefits: has to do with how much each gender group will actually benefit from the research output(s).

Participation: has to do with the freedom to get involved from design to evaluation of the research results and impacts.

Risks: has to do with anticipated risks and ability to mitigate by the gender group.

Fig. 6.2 The nine variables used for the gender-feasibility analysis. (Photos: Culture (https://www.entrepreneur.com/article/361793), Equity (https://counseling.ufl.edu/resources/bam/module1-3/), Technical competence (https://www.circularonline.co.uk/opinions/is-technical-competence-competent/), Time to benefit (https://www.dreamstime.com/illustration/benefit-opportunity.html), Cost (https://www.thedroidsonroids.com/blog/mobile-app-development-cost-in-2021), Location (https://www.indiatoday.in/information/story/how-to-correct-a-location-in-google-map-1736859-2020-10-31), Magnitude of benefits (https://pt.slideshare.net/GEMwrld/calculating-costs-and-benefits-of-investing-in-retrofitting/6), Participation (https://dailynous.com/2020/10/27/counting-participation-philosophy-classroom/), Risks (https://www.projectcentral.com/blog/project-risk-management/))

(iv) **Gender-feasibility (or quick scan) analysis**: This tool helps the researcher to identify what works and the implication of the choice of a gender group. The purpose of this analysis is to access options to meet the priority needs identified and to determine which of the identified options is most feasible based on some variables. The process quickly explores nine variables to establish whether or not the expected results of the research have gender implications. There are nine variables to consider when choosing or using this gender analysis tool (AWARD, 2014). They are shown in Fig. 6.2.

References

African Women in Agricultural Research and Development. (2014). *AWARD Science skills course participants workbook* (pp. 5–26).

Assefa, B., & de Roo, N. (2015). *Manual on gender analysis tools.* Capacity building for scaling up of evidence-based best practices in agricultural production in Ethiopia, pp. 1–42.

Gender mainstreaming. (1998). Conceptual framework, methodology and presentation of good practices. *EG-S-MS, 98*, 2.

Hosein, G., Basdeo-Gobin, T., & Gény, L. R. (2020). "Gender mainstreaming in national sustainable development planning in the Caribbean". *Studies and Perspectives series-ECLAC Subregional Headquarters for the Caribbean*, No. 87.

Leong, T. G., Lang, C., & Biasutti, M. (n.d.). *Gender analysis tools. Vibrant communities-gender and poverty project* (pp. 1–11). Canadian International Development Agency (CIDA). Retrieved 3/3/2021 from http://dmeforpeace.org/sites/default/files/CIDA_Gender%20Analysis%20Tools.pdf

Odogwu, B. A. (2020). *Wearing our gender lens in research design and development. Scientific talk series.* Accessed on the 3/3/2021 from https://owsd.net/news/news-events/owsd-nigeria-national-chapter-presents-wearing-our-gender-lens-research-design

Santiago, Economic Commission for Latin America and the Caribbean (ECLAC). (2020). UNIFEM, 1995:7.

Udry, J. R. (1994). The nature of gender. *Demography, 31*(4), 561–573.

UNESCO Gender Lens. (2003). *Project design and review.* Accessed from the UNESCO's Gender Mainstreaming Resource Center at http://www.unesco.org/women on the 20/03/2020.

Chapter 7
Impostor Syndrome with Women in Science

Rachel Paterson and Ijeoma Favour Vincent-Akpu

7.1 Introduction

The impostor syndrome (also known as impostor phenomenon, fraud syndrome, perceived fraudulence, or impostor experience) was first described by Pauline R. Clance and Suzanne A. Imes as a pervasive psychological experience of perceived intellectual and professional fraudulence (Clance & Imes, 1978; Clance, 1985a). They coined the term "impostor syndrome" after observing many high-achieving women who tended to believe they were not competent and that they were overevaluated by others. It can also be regarded as when people experience being an impostor and the thoughts and feelings elicited by such experience. The word impostor is sometimes spelt as "imposter" in some literatures. Impostor syndrome, impostor, and impostor phenomenon are the same and are used interchangeably.

Impostor syndrome was thought to occur more frequently in women than in men. This is because women are more likely to feel that an external cause (such as luck or good timing, etc.) is responsible for a success than men. Men are more likely to attribute success to their own ability and hard work. Gender tends to be insignificant with regard to the severity of impostor phenomenon. But most studies have shown that females are relatively common sufferers (Legassie et al., 2008; Maqsood et al., 2018).

This syndrome was originally identified among high-achieving professional women, but more recent researcher have documented these feelings of inadequacy among men and women in many different occupations, professional settings, from

R. Paterson
Professional Training & Coaching, San Francisco, CA, USA

I. F. Vincent-Akpu (✉)
Hydrobiology and Fisheries Unit, Department of Animal and Environmental Biology, Faculty of Science, University of Port Harcourt, Port Harcourt, Nigeria
e-mail: ijeoma.vincent-akpu@uniport.edu.ng

E. O. Nwaichi (ed.), *Science by Women*, Women in Engineering and Science, https://doi.org/10.1007/978-3-030-83032-8_7

83

different ethnic and racial groups (Hawley, 2016; Bussotti, 1990; Topping, 1983; Mak et al., 2019). Impostor syndrome arises when high-achieving individuals like women in science who, despite their successes, fail to internalize their accomplishments, competence, or skill and have persistent self-doubt and fear of being exposed as intellectual fraud or impostor. It is characterized by chronic feelings of self-doubt, fraudulent ideation, self-criticism, achievement pressure and negative emotions, and difficulty in attributing their performance to their actual competence while behaving in ways that maintain these beliefs (Vaughn et al., 2020). They attribute their successes to external factors such as luck or receiving help from others and attribute setbacks as evidence of their professional inadequacy.

Women in science are found in the fields of pure and applied science, engineering, animal and human health, and technology with highly practical tasks and experimentation. Science fields are work environments that are likely to trigger impostor syndrome than others. It is a high-achieving field which is always competitive. In addition to impostor syndrome, women in science are faced with more problems like cultural gender biases and stereotypes. Scientific revolution has done little to change people's ideas that by nature women are inferior and subordinate to men and suited to play a domestic role as nurturing mothers. The idea of women participation in science is seen as unwomanly which is prominent in some countries.

Thus, a greater understanding of what impostor syndrome is and its contributing factors and characteristics may lead to effective interventions that will reduce its consequences and psychological distress. This will reduce anxiety, increase job satisfaction, and enhance performance in workplaces.

7.2   What Impostor Syndrome Is Not

Though most people experience some kind of self-doubt when faced with new challenges as in an unfamiliar environment, new transitional experience, new breakthrough in research, or new appointment, it becomes impostor syndrome when they have persistent or chronic feelings of self-doubt and fear of being exposed as intellectual fraud or impostor. They have an all-encompassing fear of being found out not to have what it takes and have trouble putting those feelings in perspective. An estimate of 70% of people will experience at least one episode of impostor phenomenon in their lives, and this experience is not limited to people who are highly successful (Gravois, 2007).

Impostor is not a clinical/psychiatric disorder since it has not been recognized as such by the American Psychiatric Association's Diagnostic and Statistical Manual nor has it been listed as a diagnosis in the International Classification of Diseases (American Psychiatric Association, 2013; World Health Organization (WHO), 2019); but it could interfere with the psychological well-being of an individual. Though it is conventionally perceived as an ingrained personality trait, yet it is not a personality trait or mental disorder. The impostor phenomenon is not a display of false modesty and is not synonymous with having low confidence. Recently, impostor syndrome has been studied as a reaction to certain situations, stimuli, and events. It is a response experienced by different people to situations that prompt such feelings (Maqsood et al., 2018).

7.3 Types of Impostor Syndrome

Dr. Valerie Young defines five types of impostor syndrome, and most people with impostor syndrome can be in one or a mixture of these types: the perfectionist, the expert, the natural genius, the soloist, and the superhuman.

A. The Perfectionist

The perfectionist is one of the most common forms of impostor syndrome. The perfectionists want to be flawless at all times; to them 99 out of 100 is a failure. They strive to be the very best, at all cost even at the detriment of their mental health. They set impossibly high standards for themselves and feel inadequate if they cannot attain that. They beat themselves up if they forget to make one minor point in a presentation; no matter their level of success, they will never feel satisfied. Any small mistake may result to excessive worry, self-doubt, or feel of incompetency.

Everyone has unconscious rules in their head about what competent means; for perfectionist they are: "I'd never make a mistake, I should excel in everything I do." The perfectionist is primarily concerned on how the work is conducted and how it turns out, rather than on their strengths, they tend to fixate on any flaws or mistakes. Though perfectionists care about their work and want nothing but the best, they sometimes slow a team down.

Wilding (2021) has provided some questions one can ask themselves, if they are not sure if this type applies to them:

- Have you ever been accused of being a micromanager?
- Do you have great difficulty delegating? Even when you are able to do so, do you feel frustrated and disappointed in the results?
- When you miss the (insanely high) mark on something, do you accuse yourself of "not being cut out" for your job and ruminate on it for days?
- Do you feel like your work must be 100% perfect, 100% of the time?

Do not tell someone in this type to stop being a perfectionist, but they should view mistake as natural part of the process. Let them accept that there is never a "perfect time" and their work will never be 100% flawless.

B. The Expert

This is the knowledge version of the perfectionist. The experts believe that before they begin anything, they must know everything about it. They are not concerned about the quality of their work, but the quantity of their knowledge. They are always trying to learn more and are never satisfied with their level of understanding. Even when they are often highly skilled, they underrate their own expertise. They continuously seek new certifications, information, or training throughout their lives to improve their competency. This type measures their self-worth by what or how much they know or can do and not how skilled they are. They feel like a failure if they have even a tiny lack of knowledge in something.

The unconscious rule in the head of the experts about what competent means is: "I should know everything in my field." This can be demonstrated in someone not applying for a position because they do not meet all the requirements or not applying for a promotion because they never feel qualified enough. This can hold people back from going forward.

Some questions to answer that will ascertain you belong to this type (Wilding, 2021):

- Do you shy away from applying to job postings unless you meet every single educational requirement?
- Are you constantly seeking out trainings or certifications because you think you need to improve your skills in order to succeed?
- Even if you have been in your role for some time, can you relate to feeling like you still do not know "enough?"
- Do you shudder when someone says you are an expert?

There is no end to knowledge and people do not need to know everything; however, assured the expert, that there is need to be smart enough to ask questions (Neilson, 2021).

C. The Natural Genius

The natural genius thinks intelligence must be inborn or they should naturally pick up a skill. They often view themselves as an impostor if skills do not come easily to them or if they have to exert more effort to be proficient. The fact that they have to struggle to master a subject or skill or something makes them feel that they are impostor. If natural geniuses are not good enough from the start or on their first try, they might abandon that activity due to shame or embarrassment. The natural genius represents a person with impostor syndrome that not only struggles with perfectionism but also sets out to achieve lofty goals on their first attempt.

Some questions to answer that will ascertain you belong to this type (Wilding, 2021):

- Are you used to excelling without much effort?
- Do you have a track record of getting "straight A's" or "gold stars" in everything you do?
- Were you told frequently as a child that you were the "smart one" in your family or peer group?
- Do you dislike the idea of having a mentor, because you can handle things on your own?
- When you are faced with a setback, does your confidence tumble because not performing well provokes a feeling of shame?

- Do you often avoid challenges because it is so uncomfortable to try something you are not great at?

Unconscious rules in the head of the natural genius are: "I should get it right the first time, I'd always know the answer, I'd always understand what I'm reading. Neilson (2021) stated that natural geniuses have to move beyond having a fixed mindset into a growth mindset and understand that real success always takes time.

D. **The Soloist**

The soloist finds it difficult to ask others for help. If they are praised for something they received help for, they do not tend to internalize it. They think it does not count because they did not do it all on their own. A soloist feels they must always accomplish tasks independently and view themselves as failures or frauds if they ask for assistance. They believe that they can achieve anything and everything alone without help from others.

It is okay to be independent, but not to the extent that you refuse assistance so that you can prove your worth. Asking for help can equate to feelings of shame, embarrassment, incompetence, or inability to prove their own worth through their productivity. The unconscious rule in their head is: "I'd never need help." They tend to take longer time to complete tasks because they are hesitant to ask for help, and will not make a good team player.

Wilding (2021) has provided some questions one could answer, if they are not sure if this type applies to them:

- Do you firmly feel that you need to accomplish things on your own?
- "I don't need anyone's help." Does that sound like you?
- Do you frame requests in terms of the requirements of the project, rather than your needs as a person?

E. **The Superhuman**

These people feel they must excel in all the different roles they play in their life, not only in their career, but also as a parent, a partner and volunteer in the community, etc., all at the same time. The superhumans often push themselves to work harder than those around them just to prove themselves. They set high expectations that they need to measure up to. These high expectations can include taking on too much responsibility, work obligations, and family tasks. They feel the need to do it all; otherwise, they are a fraud. The superwoman/superman represents a person with impostor syndrome that often struggles with work addiction. This person may feel inadequate relative to colleagues and continue to push themselves as hard as possible, regardless of the consequences on their mental, physical, and emotional health.

Wilding (2021) has provided some questions one can ask themselves, if they are not sure if this type applies to them:

- Do you stay later at the office than the rest of your team, even past the point that you have completed that day's necessary work?

- Do you get stressed when you are not working and find downtime completely wasteful?
- Have you left your hobbies and passions fall by the wayside, sacrificed to work?
- Do you feel like you have not truly earned your title (despite numerous degrees and achievements), so you feel pressed to work harder and longer than those around you to prove your worth?

The superhumans need to be reminded that in attempting to be "the best" in every different role they play, they are not only setting themselves up to fail, but they will be stressed, overwhelmed, and burned out constantly. They need to be encouraged to slow down, set healthy boundaries, and learn how to say "no" sometimes. The superhumans need to delegate activities and allow others to participate. Being a superhuman sends an unhealthy message to those who look up to you, at home or at work.

7.4 Summary of the Simplest Way the Different Types Measure Their Accomplishments

The perfectionist is concerned on how something is done. They feel like a failure with even the smallest mistake.

The expert is concerned about what or how much they know or can do. They feel like a failure if they have a tiny lack of knowledge in something.

The soloist is concerned about "who does the thing." They feel they cannot take help from others if they want to be successful.

The natural genius is concerned about how and when accomplishments happen in terms of ease and speed. They are ashamed to take extra time or need to redo something.

The superwoman/superman is concerned about how many roles they can juggle and excel in.

Knowing which category one falls into will help the person develop the right tools to manage it. Impostor syndrome will take over if they are unable to do their job up to these standards set by the different types.

7.5 Possible Causes and Treatment of Impostor Syndrome

There is a paucity of research on the causes of the impostor phenomenon, but it could probably be attributed to a mix of nature and nurture, and cultural expectations about the role of an individual in a certain society such as upbringing, personality, and genetic factors. Investigation into the cause of impostor syndrome has been linked to certain family situations in the early childhood of an individual which are strengthened through socialization for achievement in adolescence and adulthood (Clance, 1985a; Bussotti, 1990). Sakulku and Alexander (2011) presented an overview of research into impostor syndrome, with particular focus on family achievement values and perfectionism, psychological distress, and their coping styles. Several studies have linked impostor phenomenon to family factors such as family environment, the relationship between family members and family structure (Bussotti, 1990), degree of parental care (Want & Kleitman, 2006; Sonnak & Towell, 2001), and family achievement values and personality factors (Chae et al., 1995). Some work environments such as science, engineering, politics, and banking are likely to trigger impostor syndrome than others. Though there were no data on specific treatment, duration, or improvements on any diagnostic tool used for impostor phenomenon in their review, Bravata et al. (2019) stated that literatures exist on interventions for validating patients' doubts and fears, addressing fears of failure, and providing group therapy in addition to advice on how to manage impostor symptoms.

7.6   Measurement Scale for Impostor Syndrome

Attempts have been made over the years to measure impostor syndrome with various measurement scales for clinical and research purposes (Vaughn et al., 2020). The first instrument was constructed by Harvey (1981), a 14-item scale, followed by the Clance Impostor Phenomenon Scale (CIPS) (Clance, 1985b), a 20-item scale as an improvement over the Harvey Impostor Scale. Other measurement scales include the 51-item Perceived Fraudulence Scale (Kolligian & Sternberg, 1991) and the Leary Impostor Scale, a 7-item instrument (Leary et al., 2000). Holmes et al. (1993) compared the scores of independently identified impostors and non-impostors on Harvey's IP Scale and Clance's IP Scale and suggested that Clance's IP Scale may be more sensitive and reliable instrument to measure when compared with others.

The Clance Impostor Phenomenon Scale (CIPS) was developed to help individuals determine whether or not they have impostor characteristics and is widely used to assess individuals' self-perceptions of intellectual and professional fraudulence. It is designed to assess individuals' personal feelings of incompetence and fraudulence, and assesses the phenomenon's components, such as fear of evaluation, self-doubt regarding one's own abilities, and feelings of phoniness (Yaffe, 2020). The Clance Impostor Phenomenon Scale (CIPS) contains 20 items which are divided into three theoretical dimensions: self-doubts about one's own intelligence and abilities (fake), tendency to attribute success to chance/luck (luck), and the inability to admit a good performance (discount).

These questions (Table 7.1) were asked on (a) fear of evaluation (e.g., "I avoid evaluations if possible and have a dread of others evaluating me"), (b) self-doubt regarding one's abilities (e.g., "I rarely do a project or task as well as I'd like to do it"), (c) feelings of phoniness (e.g., "I can give the impression that I'm more competent than I really am"), (d) fear of being exposed by others as a fraudster (e.g., "Sometimes I'm afraid others will discover how much knowledge or ability I really

Table 7.1 Questions in the Clance Impostor Phenomenon Scale (CIPS) (Clance, 1985b)

SN	Questions
1.	I have often succeeded on a test or task even though I was afraid that I would not do well before I undertook the task.
2.	I can give the impression that I am more competent than I really am.
3.	I avoid evaluations if possible and have a dread of others evaluating me.
4.	When people praise me for something I have accomplished, I am afraid I will not be able to live up to their expectations of me in the future.
5.	I sometimes think I obtained my present position or gained my present success because I happened to be in the right place at the right time or knew the right people.
6.	I am afraid people important to me may find out that I am not as capable as they think I am.
7.	I tend to remember the incidents in which I have not done my best more than those times I have done my best.
8.	I rarely do a project or task as well as I would like to do it.
9.	Sometimes I feel or believe that my success in my life or in my job has been the result of some kind of error.
10.	It is hard for me to accept compliments or praise about my intelligence or accomplishments.
11.	At times, I feel my success has been due to some kind of luck.
12.	I am disappointed at times in my present accomplishments and think I should have accomplished much more.
13.	Sometimes I am afraid others will discover how much knowledge or ability I really lack.
14.	I am often afraid that I may fail at a new assignment or undertaking even though I generally do well at what I attempt.
15.	When I have succeeded at something and received recognition for my accomplishments, I have doubts that I can keep repeating that success.
16.	If I receive a great deal of praise and recognition for something I have accomplished, I tend to discount the importance of what I have done.
17.	I often compare my ability to those around me and think they may be more intelligent than I am.
18.	I often worry about not succeeding with a project or examination, even though others around me have considerable confidence that I will do well.
19.	If I am going to receive a promotion or gain recognition of some kind, I hesitate to tell others until it is an accomplished fact.
20.	I feel bad and discouraged if I am not "the best" or at least "very special" in situations that involve achievement.

lack"), and (e) the inclination to underestimate self-achievements and to attribute them to external factors (e.g., "At times, I feel my success has been due to some kind of luck").

The respondents are expected to give their answer on a 5-point Likert-type scale of not at all true, rarely, sometimes, often, and very true. The total score ranges from 20 to 100: If the total score is 40 or less, the respondent has few impostor characteristics; if the score is between 41 and 60, the respondent has moderate impostor phenomenon experiences; if the score is between 61 and 80, the respondent frequently has impostor feelings; and a score higher than 80 means the respondent often has intense impostor phenomenon experiences. The higher the score, the more the impostor feeling.

Mak et al. (2019) stated that since developmental trajectory of the impostor phenomenon is currently unknown, a rigorous review of psychometric properties and
justification for the use of a specific scale is necessary, until greater understanding is established of the longitudinal variability of impostorism scores they observed for different ages.

7.7 Sufferers of Impostor Phenomenon

One may be suffering from or exhibiting signs of impostor phenomenon if one continually:

1. Worry that others may find out that they are not as smart and capable as they thought one is, or believe that other people (colleagues, competitors) are more competent than one truly is.
2. Shy away from challenges because of nagging self-doubt or do not take on challenges where you might be exposed.
3. Tend to chalk your accomplishments up to being "a fluke" or "no big deal" or to the fact that people just "like" you. When you do succeed, you think it was only a matter of being in the right place at the right time or due to chance.
4. Tend to feel crushed by even constructive criticism, seeing it as evidence of your "ineptness."
5. Live in fear of being found out, discovered, unmasked, and never put yourself in positions where you may not succeed.
6. Do all work perfectly well or overwork. Read one more book, get one more degree, or work at one more job before considering yourself qualified.
7. Always remind yourself and others of how much you really do not know.
8. Always anticipate problems before they occur.
9. Never make mistakes.
10. Do not believe compliments and praise, especially from friends and family.
11. Play down or do not take credit for your own accomplishments.

In addition, there are several feelings a person suffering from impostor phenomenon may exhibit. Do you feel:

- Like success is impossible?
- Incompetent despite demonstrating competency?
- Fear of not meeting another person's expectations?
- Like past successes and hard work were only due to luck?
- Incapable of performing at the same level every time?
- Uncomfortable with receiving praise or congratulations?
- Disappointed over current accomplishments?
- Doubtful of successes?

- Constant pressure to achieve or be better than before?
- Stressed, anxious, or depressed from feelings of inadequacy?

If you have affirmative responses to some of the actions listed above, you are a sufferer of impostor phenomenon or exhibiting signs of impostor phenomenon.

7.8   Potential Characteristics of Impostor Phenomenon

Clance (1985a) suggested six potential characteristics of impostor phenomenon:

1. The impostor cycle.
2. The need to be special or to be the very best.
3. Superman/superwoman aspects.
4. Fear of failure.
5. Denial of competence and discounting praise.
6. Fear and guilt about success.

The impostor cycle starts when an important task is assigned to an individual with impostor phenomenon. It begins with an initial anxiety which they react by extreme over-preparation or initial procrastination followed by hurried preparation. When the task is done, there is an initial sense of accomplishment and relief, which do not last despite receiving positive feedback about their accomplishment. They deny that their success is due to their own ability. When faced with a new task, self-doubt creates a high level of anxiety, and the impostor cycle is repeated. The impostor cycle is one of the most important characteristics of the impostor phenomenon and is difficult to break (Clance, 1985a). Impostors often have the feeling of wanting to be the best when compared with their peers, maybe the top of the class at smaller settings. However, in a larger setting, such as in a university, impostors realize that there are many others better than them and as a result they feel overwhelmed or disappointed, and consider themselves as failures when they are not the very best.

Impostors tend to do everything flawlessly in every aspect of their lives. Clance (1985a) stated that "the need to be the very best" and "the superman/superwomen aspects" are inter-related which she refers to as perfectionistic tendency. They set high standards and often feel they have failed when they are unable to fulfill their perfectionistic goals (Clance, 1985a).

Impostors experience high levels of anxiety when exposed to an achievement-related task because they fear possible failure. The fear of failure or making mistakes is an underlying motive of most impostors. To reduce the risk of possible failure, impostors tend to overwork. Impostors reject their competence and find it difficult to accept praise as valid. They attribute their success to external factors to a greater degree than non-impostors (Chae et al., 1995; Harvey, 1981). They discount positive feedback and objective evidence of success but focus on evidence or develop arguments to prove that they do not deserve praise or credit for particular achievements (Clance, 1985a).

Impostors have fear or guilt about their success especially if their successes are unusual in their family or their peers which often make them feel less connected and more distant. They are frightened that their success may lead to higher demands and greater expectations from people around them and worry about being rejected by others (Clance, 1985a).

These Clance's six potential characteristics vary in impostors. Not every impostor experiences all these characteristics, but to be considered as an impostor, a minimum of two characteristics should be found. Harvey and Katz (1985) proposed that the impostor phenomenon consists of three core factors: (1) the belief of having fooled other people, (2) fear of being exposed as an impostor, and (3) inability to attribute their achievement to internal qualities such as competence, intelligence, or skills. All these three criteria must be met in order to consider someone an impostor.

7.9 Consequences of Impostor Phenomenon

Studies have shown that impostor syndrome has shown strong correlations with measures of depressive thoughts and feelings and characteristics of depression. The feeling of impostor syndrome, if prolonged, may lead to clinical levels of depression or some aspects of psychological distress. They suffer from anxiety, fear of failure, and dissatisfaction with life and believe themselves to be less intelligent and competent than others perceive them to be. Impostor syndrome is associated with low level of self-esteem, being easily discouraged, obsessed over mistakes and failures, lack of confidence, and repetitive negative self-talk. The impact of these feelings and thoughts can lead to struggling to accept praise and recognition, internalizing all flaws, mistakes, and criticisms.

There is a continuous need to be the best, to do everything flawlessly, and judging themselves by only the highest standards or perfectionism. This tends to lead to overworking, holding back, and procrastination. They might stay in a particular job they have long outgrown because they are afraid to pop their head up or being found out as a fraud." The perfectionistic expectations of impostors have contributed to the feeling of inadequacy, increasing levels of distress, and depression. All of these happen when impostors perceive that they are unable to meet the standards they set for themselves or expectations from family and people around them (Clance, 1985a). They set high or unrealistic goals and then experience a sense of self-defeat "when they cannot reach those goals" (Kets de Vries, 2005). However, impostors have higher level of achievement motivation than non-impostors in order to eradicate their own personal sense of self-doubt and prove they are capable, competent, and worthwhile (Topping, 1983). Impostors feel uncertain about their ability to maintain their current level of performance and are reluctant to accept additional responsibility.

7.10   How to Harness the Inner Impostor

Having confidence in the quality of your work is a basic necessity for someone to be effective in a profession in science. Feeling like a phony will undermine the quality of your work. In order to restore your positive feelings about yourself, you must first explore the reasons why you have viewed yourself as an impostor and then address those reasons so you can regain your confidence!

There are four self-assessment questions that can help you discover what might be making you uneasy:

1. What are the roots of your uncertainty?
2. If you have just taken on a new position, a new role, and some new research, how prepared do you feel?
3. Are you convinced that "luck" is the only reason you are advancing in your field?
4. How much of a perfectionist are you and have you set goals that are unachievable?

Larsen (2016) laid out six simple but effective steps to harness your inner impostor which has been grouped into three steps: awareness, connection, and harnessing steps.

A. Awareness Steps:

  1. Understand it is a common occurrence and you are in good company since up to 75% of successful leaders and entrepreneurs have admitted to experiencing impostor syndrome at some point in their career.
  2. Understand that it is originating from you and you are the author of it; it is not coming from other sources or other people.

B. Connection Steps:

  3. Connect to your past accomplishments, your past triumphs, your past learnings, or your past legacy. Write your success story, or what has gotten you to this place. What you have achieved is not luck or fate, but deliberate steps you have taken that have made you who you are.
  4. Connect to your future vision, your future purpose, your future goals/outcomes, or your future potential. Author a compelling future vision of yourself by painting a realistic picture of what success would look like, feel like, etc. (Impostor syndrome thinks of the present hence "why we want to connect the past and future" so that we do not ignore the present but use it as a stepping stone from our past to our future.)
  5. Harness your past and future legacies to create your three anchors of your present success. Create a triangle with three points and label each point with an attribute/trait/skill that you do well, that you are recognized for. What makes you successful in the present?

C. Harnessing Steps:

  6 Harness your three anchors to continue to move forward, get unstuck, finesse
  your triggers when they arise, and use it all for building a confident and convincing leadership brand. Use your three anchors to bring awareness of what
  you have done in the past to be successful, what you do in the present to
  remain successful, and what you need to do in the future to be successful.

7.11 Tips for Harnessing Impostor Syndrome (Cuncic, 2021; Sobara, 2020)

1. Ask the difficult questions: There is no room for women to be meager and shy in
   science. You do not hold back when you have a question; ask it without questioning how others may feel. It is not because you are not innovative but you are
   unwilling to speak up or hesitant to share your ideas. Women in science have to
   go the extra mile to prove their worth; they must speak up and speak out about
   what they are passionate about.
2. Find support and be supportive of others: Talk to other people about how you are
   feeling and you will find out you are not alone. These impostor feelings tend to
   fester when they are hidden and not talked about. Determine your support system
   and lean on them when needed. Discuss feelings of inadequacy with others (e.g.,
   with friends and family or at individual or group therapy). A better strategy is for
   women scientists to bind together and support each other in their pursuit of
   STEM careers. There are many nonprofit organizations for women in science
   that can help to build your network and provide mentorship along the way. Seek
   such out.
3. Fail forward: Remember that nobody is perfect; best accomplishments are
   achieved after many rounds of trial and error. It is a necessary part of scientific
   experimentation and is the same in professional development. Even the most
   brilliant minds have made mistakes. We do not expect absolute perfection from
   others at all times and should not demand absolute perfection from ourselves at
   all times. Allow yourself the lenience to make mistakes and learn from them. As
   long as we learn from our mistakes, we will continue to grow and become stronger and more intelligent. Set clear, measurable, and realistic goals and set limits
   or boundaries to avoid overworking.
4. Wear many hats: As a woman, you are not defined by any one job or any one
   thing. Too many women confine themselves to a single pursuit. They falsely
   believe they can only be scientists or businesswomen or mothers. In reality,
   women can be many things at once. We do not have to fit ourselves into a box.
   There is no perfect job, and the women who go on to be successful are those who
   are willing to wear many hats and refuse to fit into any one mold.
5. Stop apologizing: Constantly apologizing for oneself is a habit not exclusive to
   women, but it is a very prominent and problematic characteristic nonetheless.

When you constantly apologize, you communicate to both yourself and the outside world that you are always wrong. This hurts both your self-esteem and your integrity.

A study published in the *European Journal of Social Psychology* shows that refusing to apologize provides several psychological benefits, including empowerment, confidence, and greater feelings of integrity and self-respect. Of course, you should apologize if you did something legitimately wrong or failed to deliver, own up to it, learn from it, and move on. However, please do not apologize for your beliefs, your desires, your goals, your past, or the fact that you are a woman; you must stop apologizing if you want to move forward.

6. Celebrate the small victories: While woman face many difficulties in the workplace, it is important to know that you do not have to be a martyr. You do not have to carry the torch for all women. Instead, all you have to do is fight your own battles and do what is best for you and your career. By doing this, you will set an amazing example for women around you and for the women scientists who will follow you.

   Stop waiting for your peers and superiors to take notice and start validating your own victories. Define what success means to you without including the approval of others. An important part of staying motivated is celebrating even the smallest victories in your career. Praise yourself for successes and efforts. At the end of each day, review your accomplishments and acknowledge that you are one day closer to achieving your career goals.

7. Embrace your self-worth: Women scientists have many advantages over other job candidates. If you are a woman and have a PhD or are on your way to having one, the future is yours. The only thing that can hold you back is yourself. Know your value and recognize your expertise. Women scientists are desperately needed in all industries, but you have to step up and seize the position you want.

8. Stop fighting your feelings: Acknowledge the impostor feelings of not belonging and start to unravel those core beliefs that are holding you back. Refuse to let it hold you back. Assess your abilities and question whether your thoughts are rational and replace them with positive thinking. Does it make sense that you are a fraud, given everything that you know? Questioning negative thoughts and no matter how much you feel like you do not belong, do not let that stop you from pursuing your goals. Keep going and refuse to be stopped.

7.12   Conclusion

To break out of impostor syndrome, women in science must identify the symptoms. These symptoms include self-doubting thoughts of not being good enough, feelings of inadequacy, being easily discouraged, believing that others are smarter, and negative self-talk, among others. When they speak up and share their feelings, they will find out that they are not alone. Sometimes, people feel stupid, that does not mean they are; or feel off-base when starting something new. They have to separate feelings from fact. Learn to ask for help. Do not be such a perfectionist; forgive yourself when you make mistake. The only way to stop feeling like an impostor is to stop thinking like one (Young, 2011). Consciously counter negative thoughts with positive statements and emphasis on your actual competencies. If the problem persists, it is important to seek for assistance from a mental health professional.

References

American Psychiatric Association. (2013). *Diagnostic and statistical manual of mental disorders* (5th ed.). American Psychiatric Publishing. https://doi.org/10.1176/appi.books.9780890425596

Bravata, D. M. A., Watts, S., Keefer, A. L., Madhusudhan, D. K., Taylor, K. T., Clark, D. M., Nelson, R. S., Cokley, K. O., & Hagg, H. K. (2019). Impostor syndrome: A systematic review. *Journal of General Internal Medicine, 35*(4), 1252–1275. https://doi.org/10.1007/s11606-019-05364-1

Bussotti, C. (1990). The impostor phenomenon: Family roles and environment. (Doctoral dissertation, Georgia State University). *Dissertation Abstracts International, 51,* 4041B–4042B.

Chae, J. H., Piedmont, R. L., Estadt, B. K., & Wicks, R. J. (1995). Personological evaluation of Clance's impostor phenomenon scale in a Korean sample. *Journal of Personality Assessment, 65*(3), 468–485.

Clance, P. R. (1985a). *The impostor phenomenon: Overcoming the fear that haunts your success* Atlanta. Peachtree Publishers.

Clance, P. R. (1985b). https://paulineroseclance.com/pdf/IPTestandscoring.pdf

Clance, P. R., & Imes, S. A. (1978). The impostor phenomenon in high achieving women: Dynamics and therapeutic intervention. *Psychotherapy: Theory, Research, and Practice, 15*(3), 241–247. https://doi.org/10.1037/h0086006

Cuncic, A. (2021). *What is imposter syndrome?* Verywellmind. https://www.verywellmind.com/imposter-syndrome-and-social-anxiety-disorder-4156469

Gravois, J. (2007). You're not fooling anyone. *The Chronicle of Higher Education, 54*(11), A1. Retrieved March 2020, Available from http://chronicle.com

Harvey, J. C. (1981). The impostor phenomenon an achievement: A failure to internalize success (Doctoral dissertation, Temple University). *Dissertation Abstracts International, 42,* 4969B.

Harvey, J. C., & Katz, C. (1985). *If I'm so successful, why do I feel like a fake?* Random House.

Hawley, K. (2016). *Feeling a fraud? It's not your fault! We can all work together against Impostor Syndrome*. Retrieved March 2020. Available from: https://www.psychologytoday.com/us/blog/trust/201607/feeling-fraud-its-not-your-fault.

Holmes, S. W., Kertay, L., Adamson, L. B., Holland, C. L., & Clance, P. R. (1993). Measuring the impostor phenomenon: A comparison of Clance's IP scale and Harvey's I-P scale. *Journal of Personality Assessment, 60*(1), 48–59. https://doi.org/10.1207/s15327752jpa6001_3

Kets de Vries, M. (2005). The dangers of feeling like a fake. *Harvard Business Review, 83*(9), 110–116.

Kolligian, J., & Sternberg, R. J. (1991). Perceived fraudulence in young adults: Is there an "impostor syndrome"? *Journal of Personality Assessment, 56,* 308–326. https://doi.org/10.1207/s15327752jpa5602_1

Larsen, P. N. (2016). *Find your voice as a leader*. Retrieved March 2020. Available from: https://paulnlarsen.com/5-steps-for-finding-your-voice-as-a-leader/

Leary, M. R., Patton, K. M., Orlando, A. E., & Funk, W. (2000). The impostor phenomenon: Self-perceptions, reflected appraisals, and interpersonal strategies. *Journal of Personality, 68*(4), 725–756.

Legassie, J., Zibrowski, E. M., & Goldszmidt, M. A. (2008). Measuring resident well-being: Impostorism and burnout syndrome in residency. *Journal of General Internal Medicine, 23*(7), 1090–1094.

Mak, K. K. L., Kleitman, S., & Abbott, M. J. (2019). Impostor phenomenon measurement scales: A systematic review. *Frontiers in Psychology, 10*(671). https://doi.org/10.3389/fpsyg.2019.00671

Maqsood, H., Shakeel, H. A., Hussain, H., Khan, A. R., Ali, B., Ishaq, A., & Shah, S. A. Y. (2018). The descriptive study of impostor syndrome in medical students. *International Journal of Research in Medical Sciences, 6*(10), 3431–3434. www.msjonline.org

Neilson, K. (2021). *Five types of impostor syndrome (and how to manage them)*. Leadership development, Strategic HR. https://www.hrmonline.com.au/section/strategic-hr/five-types-imposter-syndrome-how-to-manage-them

Sakulku, J., & Alexander, J. (2011). The impostor phenomenon. *International Journal of Behavioral Science, 6*(1), 75–97.

Sobara, C. (2020). *Tips to help women scientists overcome impostor syndrome and transition into industry*. Retrieved March 2020. Available from: https://cheekyscientist.com/7-tips-to-help-women-scientists-overcome-impostor-syndrome

Sonnak, C., & Towell, T. (2001). The impostor phenomenon in British university students: Relationships between self-esteem, mental health, parental rearing style and socioeconomic status. *Personality and Individual Differences, 31*(6), 863–874.

Topping, M. E. (1983). The impostor phenomenon: A study of its construct and incidence in university faculty members. (Doctoral dissertation, University of South Florida). *Dissertation Abstracts International, 44*, 1948B–1949B.

Vaughn, A. R., Taasoobshirazi, G., & Johnson, M. L. (2020). Impostor phenomenon and motivation: Women in higher education. *Studies in Higher Education, 45*(4), 780–795. https://doi.org/10.1080/03075079.2019.1568976

Want, J., & Kleitman, S. (2006). Feeling "phony": Adult achievement behaviour, parental rearing style and self-confidence. *Journal of Personality and Individual Differences, 40*(5), 961–971.

Wilding, M. J. (2021). *5 different types of Imposter Syndrome (and 5 ways to battle each one)*. Themuse https://www.themuse.com/advice/5-different-types-of-imposter-syndrome-and-5-ways-to-battle-each-one#:~:text=Valerie%20Young%2C%20has%20categorized%20it,in%20Spite%20of%20It%2C%20Dr

World Health Organization (WHO). (2019). *International statistical classification of diseases and related health problems 10th revision (ICD-10)-WHO version for; 2019-covid-expanded*. https://icd.who.int/browse10/2019/en#/V

Yaffe, Y. (2020). Validation of the Clance impostor phenomenon scale with female Hebrew-speaking students. *Journal of Experimental Psychopathology*, 1–8. https://journals.sagepub.com/doi/pdf/10.1177/2043808720974341

Young, V. (2011). *The secret thoughts of successful women: Why capable people suffer from the imposter syndrome and how to thrive in spite of it*. Currency. 46773rd ed., pp. 304.

Chapter 8
The Beauty of Research Data in an Information-Driven World

Linda Uchenna Oghenekaro

8.1 Introduction

Data has become one of the most valuable assets in recent times, and research data, in particular, has proven to be an important resource that fuels the knowledge economy. The knowledge economy can be understood to be an economic system, where goods and services are produced based on intensive knowledge activities that, in turn, rapidly advance scientific and technical innovations. As it is rightly said by Clive Humby, "Data is the new oil in today's information economy." An information-driven economy sources data from multiple locations, in different languages and formats, for an absolute representation of the problem at hand. It is worth noting that an information-driven economy differs from an economy driven by information, as the latter is one-off, while the former is continuous. Science, as we know, is a data-driven field, and the quality of our scientific study depends largely on the value of our research data. Fortunately, the swift advancement in computing technology has birthed new applications for sourcing for research data, and this development has given a major thrust to scientific research in recent years. Effective access to research data, in a responsible and efficient manner, is required to take full advantage of the new opportunities and benefits offered by information and communication technologies.

L. U. Oghenekaro (✉)
Department of Computer Science, Faculty of Science, University of Port Harcourt, Port Harcourt, Rivers State, Nigeria
e-mail: linda.oghenekaro@uniport.edu.ng

E. O. Nwaichi (ed.), *Science by Women*, Women in Engineering and Science, https://doi.org/10.1007/978-3-030-83032-8_8

8.2   What Is Research Data?

It remains a challenging task to give a singular definition of what a research data is, because it varies across different disciplines and research funders, research data can be basically defined as recorded facts or statistics generated and collected for processing and interpretation in a bid to produce original research results.

8.3   Forms of Research Data

Depending on the type of research, we see research data take up several forms, which are discussed as follows:

1. Documents: Research data in this form can be organized in individual words, connotations, or whole sentences; in general, their appearance can be traced to text form.
2. Spreadsheets: This form captures data precise numerical values, using spreadsheet applications. The absence or presence of specific characteristics within the data is highlighted in this form.
3. Sensor readings: This form of research data is gotten from automatic data acquisition systems that extract great amount of streaming data generated by physical sensors.
4. Laboratory notebooks: This is maintained for researches done in the laboratory, and it contains information like dates, reagent data, details of experiments and procedures, methods, observations, measurements, and continuation notes.
5. Scripts: This form of research data is generated by scripting languages in computational researches and disciplines. Varieties of tools are implemented in several scripting languages, using high-performance computing systems (Wang and Peng, 2019).
6. Questionnaires: They are made up of standardized questions, and this form of research data is seen and adopted in survey research. They follow a fixed scheme in collecting individual data on specific topics.
7. Photographs: These are made up of visual images that encompass different artifacts that can be applied in qualitative research methodologies.
8. Audio and video tapes: Some subject disciplines adopt this form of research data to capture procedures, processes, and interactions.

8.4   Sources of Research Data

The sources of research data serve as the base material of every research. A good understanding of data source helps the researcher to be informed on what approach should be applied in collecting the data for use. Research data can basically be classified into two broad types, primary and secondary sources, as illustrated in Fig. 8.1.

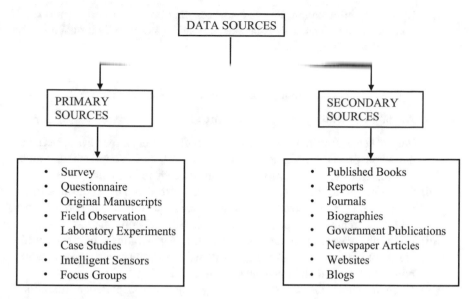

Fig. 8.1 Types of research data sources

Primary Source

A primary source of scientific research data provides firsthand evidence about the research area of interest; it is usually carried out in either a controlled or uncontrolled situation, with the end goal of the research in mind. A case of controlled environment is when certain variables are controlled by the researcher or experimenter. Primary sources always present novel reports on discoveries and new information. Data generated from this source is unique and original and can be considered as primary data. The methods used to generate data from primary source include experimental, survey, or observation methods.

A Advantages of Primary Source of Research Data.

   (i) Resolution of specific research issues: Primary source holds specific information that is exact to what the researcher desires, and it is tailored to specific research needs. The researcher also reports his/her findings in a way that benefits the purpose of the research.

  (ii) Better accuracy: Given that data is sourced directly from the population of interest, there is a high possibility of getting more accurate data.

 (iii) Primary ownership: Primary sources of research data are fully owned by the researcher or group of researchers that generated the original data and, hence, have control over the management and access of their data.

(iv) Up-to-date information: Primary sources play host to up-to-date information on research data. This is because researchers collect their data directly from the field in real time.

(v) High level of control: Control can be achieved at different points in primarily sourcing for research data; from gathering data to the point of publishing, researchers have high level of control over the entire process.

B Disadvantages of Primary Source of Research Data.

(i) Very expensive: Primary source of data could be very expensive, given that the whole procedure begins from the very start, which implies that collection materials and other activities involved needs to be funded.

(ii) Consumes time: Building a primary source of research data consumes a lot of time, and for some primary sources, time is a priority factor, and for some other researches, the data generated depends largely on time factor.

(iii) Impossible cases: There are researches where generating a primary source is almost impossible, as a result of certain factors such as large size of target population needed and financial cost involved.

(iv) Limited factors: Generation of primary source can experience limitations such as complex nature of the work, location, time, and staff strength, among other factors, and these limitations become bottlenecks to creating a primary source of research data.

Secondary Source

This source holds data that has been collected in the past. It can be further divided into internal and external sources. The internal source refers to incorporated data that exists within the researchers' organization, while the external source refers to gathering data from associations outside one's organization. The data gotten from secondary source have already been retrieved in the past. Secondary sources of data include government record, newspaper, journals, repositories, books, websites, etc. The use of the Internet has made secondary sources of data more accessible. Data from secondary sources can be in electronic or printed form; they may also be free or paid, depending on the publishing company of the primary researcher's decision.

A Advantages of Secondary Source of Research Data.

(i) Inexpensive: Most secondary sources of research data are available online for free download. However, books and research materials can also be accessed for free from public libraries, for researchers who do not have adequate access to the Internet.

(ii) Time-saving: The time involved in accessing secondary source is usually little as compared to primary source, thereby saving a significant amount of time for the researcher. Secondary source is considered to be readily available in most cases.

(iii) Ease of comparative and longitudinal studies: With secondary source, researchers can carry out longitudinal studies without necessarily waiting for a number of years to get data or draw conclusions. Comparative studies ⬛⬛⬛⬛⬛⬛⬛⬛⬛⬛⬛⬛⬛⬛⬛⬛⬛⬛⬛⬛⬛⬛⬛⬛⬛⬛⬛⬛⬛⬛⬛⬛⬛⬛⬛ ent attributes of interest, such as time.

(iv) Generation of new insights: New insights can be gotten from secondary sources, as data collected in this source can be analyzed from another point of view, leading to new discoveries that were probably not revealed by the primary data collector.

B Disadvantages of Secondary Source of Research Data.

(i) Data quality: Some secondary sources are not as authentic as the primary source. And this is a very common negative aspect with online secondary source because of the lack of monitoring and regulatory bodies for the content being shared.

(ii) Exaggerated data: This drawback is common to secondary sources such as blogs, where certain data are exaggerated for personal reasons, such as; to gain web traffic, public attention, attract customers, or even satisfy a paid advert.

(iii) Outdated information: Some secondary sources still maintain outdated information, rather than replacing them with new ones. For example, a government online site that is expected to publish its report on a yearly basis, but has a 5-year-old data as its most recent report, will be forcing researchers to make do of an outdated data for a research.

(iv) Irrelevant data: Secondary sources play host to a number of irrelevant data, and the reason is because the data was not collected primarily for the researcher.

Although secondary sources of research data have their drawback, which poses negative effects on the general outcome of a research, they also have some advantages over the primary source. The choice of research data source to adopt depends on the researcher and the nature of research being carried out. It is worth noting that there are cases where secondary source may be the only source of data, cases such as access to data being delegated to a particular body, or huge cost of getting the data, firsthand. Figure 8.1 illustrates the different types of data sources.

8.5   Categories of Research Data

Depending on the nature of the research, a choice can be from the two broad categories of research data.

Quantitative Data

They are measurable data used to formulate facts, and they are structured, numerical in nature, discrete, and statistical. Researchers adopt them when they intend to quantify a research question or a problem. Quantitative data can either be compared on a numerical scale or counted; they are measured objectively. Examples of quantitative data are data gotten from measurements, for instance, the distance, measured in kilometers, between a given community and a public market within a local government area.

Qualitative Data

This category holds text-based data where it describes qualities and characteristics; it is less structured in nature. They are usually gotten from questionnaires and interviews and appear in a narrative form; they are observed subjectively. It is more concerned in describing a topic rather than measuring it. For example, responses from an open-ended questionnaire on the preferred route that indigenes of a given-community would take to access their public market.

Fortunately, both categories do not contradict each other; rather, they complement themselves. A research might begin with the quantitative data to prove the general points of the study and then adopt the qualitative data to give more details of one's findings. Figure 8.2 illustrates categories of research data, and a further classification can be seen where the quantitative is further divided into continuous and discrete data; continuous data are measured within a certain range, and discrete data are measured at exact values and counts. The qualitative data, which is non-numerical, is seen to be further divided into nominal and ordinal data, where the nominal refers to qualitative data that has no specific order and ordinal suggests specific order. Table 8.1 highlights the cost and benefits of the two broad categories of research data.

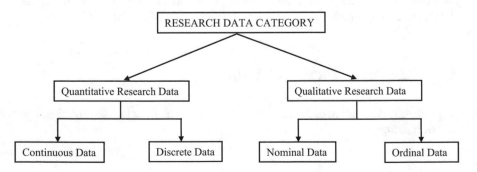

Fig. 8.2 Categories of research data

Table 8.1 Category of research data trade-offs

Research data category	Advantages	Disadvantages
Quantitative	Reproducible knowledge can be generated Can systematically represent a large collection of data	Large samples required Statistical training may be needed to analyze data
Qualitative	Methods can be adjusted as new knowledge is developed Can be conducted with small samples	Difficult to standardize research Cannot be generalized to broader populations Cannot be analyzed statistically

8.6   Research Data Collection Techniques

Research data can be generated or collected by the researcher using either the primary or secondary data collection techniques. For primary data collection, techniques such as experiment, survey, or direct observation can be applied, while techniques used in collecting information from existing sources such as data repositories, market surveys, and standard reports can be applied for secondary data collection.

Primary Data Collection Techniques

(a) Experiment: The researcher generates firsthand data from structured study in a controlled environment, to understand the cause and effect of a given process and record all findings accordingly. This technique is usually carried out in the laboratory. Its strength lies in its objective nature of data recording; however, incorrect data could still be recorded as a result of human error.

(b) Interview: This can be carried out in person or over the phone, and in whichever way, the researcher is the interviewer who asks the questions, and respondent is the interviewee who provides the answers. The strength of interviews is that comprehensive information can be collected and bias responses can be detected. However, it is more time consuming and expensive.

(c) Survey: This can also be likened to a questionnaire, where a group of questions is written down or typed and sent to a target sample of study for their responses. The researcher records these responses as original data. Before questionnaires are sent out, it is advisable to present them to experts to assess the strength and weakness of techniques and questions used. This technique is cheaper than interviews, and respondents have ample time to give responses; however, it is a slower process as compared to interviews, and it is also not flexible, as responses cannot be changed once submitted.

(d) Direct observation: This technique is mostly adopted in researches related to behavioral science and can be carried out in various approaches such as structured or unstructured, controlled or uncontrolled, with or without participants. Data gotten for this technique is usually objective, as it is not affected by past or future actions. However, information gotten via direct observation can be limited.

Secondary Data Collection Techniques

In all cases, secondary data are already existing data, either published or unpublished. Hence, collection of this class of data is straightforward, by getting the right access to the respective repository. The authenticity of the collected data should also be verified, so as not to reduce the overall quality of the research work.

8.7   Guidelines for Generating Research Data

(a) Start sourcing for your data early, to avoid falsifying them.
(b) Avoid factors that will lead to bias; be open to your findings on the field.
(c) Consider large sample size, as it increases the accuracy of our research results.
(d) Consult credible sources, for secondary data collection.

8.8   Research Data Life Cycle

The life cycle of research data provides researchers with a high-level overview of the stages involved preserving data for use and reuse (Wang and Peng, 2019). It generally has about seven stages, where every individual stage is made up of several activities that are carried out to achieve the purpose of the given stage. Figure 8.3 gives an illustration of the stages:

  (i) Plan: The researcher identifies the data to be generated, which should be able to address the research question, and then plans on how the data will be managed throughout its life cycle.
 (ii) Collect: At this stage, the researcher adopts an appropriate data collection technique, either primary or secondary, to generating research data, depending on the nature of the research. This stage involves documentation of the data collection method used, instruments, and every activity carried out in generating the data.
(iii) Process: The collected data undergoes processing at the stage, which might involve cleaning the data of noise, transforming data from one format to another, and combining data from multiple sources, and all processes must be documented for proper referencing.

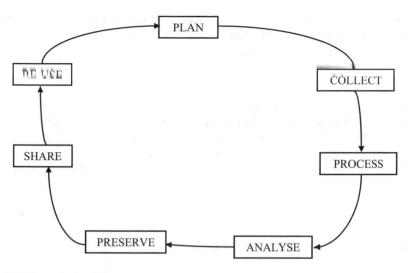

Fig. 8.3 Research data life cycle

(iv) Analyze: The processed data is analyzed at the point to produce insights that form the research findings, which are either published in research outputs or simply documented. Materials and methods used in data analysis should also be documented.

(v) Preserve: Activities in this stage happens towards the end of a research work. In this stage, data is prepared for preservation and it is deposited or archived in a suitable repository. Activities in this stage include creation of metadata with the assignment of DOI (Digital Object Identifiers) to the datasets, ensuring access control, and licensing data for reuse, among others.

(vi) Share: Preserved data can either be made publicly available or restricted until access is applied for by the interested parties. A data repository enables data in its care to be discovered, by displaying its metadata online, and providing data when the necessary access is granted.

(vii) Reuse: Data may be reused by other researchers, as a secondary source, to either generate new findings through further analysis or authenticate the findings of the original research (Whyte and Tedds, 2011). At this stage, the data can be considered as raw material for a new research cycle. Research data can be further used to make policies, to teach, and also to develop products and services in both industry and academia.

8.9   Research Data Management

Given that research data has a longer lifespan than the research itself, there is a need to properly manage our data through its life cycle. Managing research data involves organizing data from the point of generation to the point of dissemination and documentation (Dunie, 2017). It has several benefits, not only to the primary researcher but also to other researchers within our domain of study.

Steps involved in managing research data include:

1. Identify the purpose for which the data is sourced.
2. Organize data as they grow in size.
3. Create a metadata for sourced data.
4. Label data files properly for easy identification.
5. Always back up research data either online or offline.
6. Use the right tools to analyze every research data.
7. Secure your data; it is an asset: Always use a combination of the Windows key and the "L" key to quickly lock your system, if the need arises.

The Need to Manage Research Data

1. Increase transparency and accessibility.
2. Enable reusability.
3. Reduce risk of data loss.
4. Improve the research integrity of home institution.
5. Avoid repetition of work, as a result of senseless collation of data.
6. Easily identify different versions of a given data.
7. Increase efficiency in report writing.
8. Enable collaboration.
9. Ease of retrieval in personal files.

8.10   Why Research Data?

Research data continues to play important roles in scientific study, and a few of its advantages have been highlighted:

1. Improves the quality of human life in general.
2. Transforms to well-informed knowledge.
3. Provides indisputable evidence.
4. Proffers solutions to problems.
5. Reduces uncertainty in decision and policy making.
6. An objective way to keep track of progress in academic and industry.
7. Serves as a catalyst for societal change.
8. Influences the quality of our research.

8.11   Conclusion

This chapter was intended to increase the level of awareness of how research data contributes significantly to the overall quality of research work. Research data was also discussed on several subtopics as it concerns our information-driven world. Research data is precious to scientific study, and as Daniel Keys Moran rightly said, "You can have data without information, but you cannot have information without data." So, as female researchers, we are encouraged to pay more attention to our research data, as this will enable us to make cutting-edge impact in our scientific world.

References

Dunie, M. (2017). The importance of research data management: The value of electronic laboratory notebooks in the management of data integrity and data availability. *Information Services and Use, 37*, 355–359.

Wang, G., & Peng, B. (2019). A pragmatic workflow system of daily computational research. *PLoS Computational Biology, 15*(2), e1006843.

Whyte, A., & Tedds, J. (2011). *Making the case for research data management*. Digital Curation Centre. Briefing Papers.

Chapter 9
Women in Academia: Developing Self-Confidence and Assertiveness

Blessing Minaopunye Onyegeme-Okerenta

9.1 Introduction

Academia describes the life, environment, community, society or world of teachers, schools and education. It is a world of research and innovation, filled with bright minds formulating and testing new theories, searching for patterns and explanations for the world around us: a field defined by its drive to push the boundaries of knowledge and characterized by progress and enlightenment. Its context includes the nursery, primary, secondary and tertiary educational community. Academia plays out most in the university system which is the tertiary and apex of education. It is a very competitive environment that is male-dominated and gender-biased.

Women not only face systemic challenges in their academic careers, but they also have to fight inherent socially constructed gender biases from their students and colleagues. A census of women in academia who are occupying the hardship position of their institution all over the world will reveal the level of marginalization the women in academia face. The belief that it is a man's world has given the men a very wrong assumption that women should play second fiddle in the academic environment since their primary role is in the home and nursery. Some women have suffered intimidation, abuse or physical harassment because they were perceived as a threat to the opposite gender and will not be voted for when they vie for a leadership position in their institution. The most challenging roles and tasks are not given to women because 'she is a woman, how can she be our boss and be calling the shots?' No wonder the outcry from women against gender discrimination and the advocacy for gender equality and women's liberation in all walks of life. In addition to facing systemic challenges in their academic careers, women also have to fight inherent

B. M. Onyegeme-Okerenta (✉)
Department of Biochemistry, University of Port Harcourt Choba,
Port Harcourt, Rivers State, Nigeria
e-mail: blessing.onyegemeokerenta@uniport.edu.ng

© The Author(s), under exclusive license to Springer Nature
Switzerland AG 2022
E. O. Nwaichi (ed.), *Science by Women*, Women in Engineering and Science,
https://doi.org/10.1007/978-3-030-83032-8_9

111

cultural and socially constructed gender norms; they are expected to have certain characteristics and behaviours (such as being respectful and caring). If they do not, society tends to judge them negatively or label them as insolent fellows (Sandberg, 2013). Women are expected at board meetings to serve tea to their male counterparts, even though they are of the same academic cadre. Often they are reminded that their roles include hospitality and that there is nothing they can do about it.

Few women are in the top echelon of the university organization, not because they are not brilliant with the basic required qualifications but because they have lost confidence in themselves or the system and have accepted that gender bias and inequality is an age-long tradition. This lack of self-confidence cuts across all facets of life and career and has impacted negatively on the mental and emotional state of the womenfolk. Some women find it difficult to publish their scholarly articles or make presentations of the outcome of their research findings because they do not believe in themselves or cannot go public due to perceived or imaginary reactions from the academic world. Others who attempted to be visible were ridiculed and reminded that a woman's place is in the kitchen and they had to put up a fight before they get any attention from their male colleagues. It is a system where the men exhibit a sense of entitlement to intimidate the women. This write-up is aimed at addressing how women in academia can develop self-confidence and assertiveness.

9.2 What Are Self-Confidence and Assertiveness?

Self-Confidence

Confidence is the quality of being certain of your abilities or of having trust in people, plans or the future; it is the key to success, peace of mind and well-being. Confidence is, in part, a result of how we have been brought up and how we have been taught. We learn from others how to think about ourselves and how to behave – these lessons affect what we believe about ourselves and other people. Confidence is also a result of our experiences and how we have learned to react to different situations. Self-confidence is all about having faith in your ability. It is an attitude or perception about your skills and abilities; it means you accept and trust yourself and have a sense of control in your life. By trusting your ability, you know your strengths and weaknesses well and have a positive view of yourself. You set realistic expectations and goals, communicate assertively, and can handle criticism. It is not something that is genetically acquired; rather, it is a phenotypic expression and a behaviour or a way of approaching things that can be learned from your environment. Having self-confidence boosts your self-esteem and helps you achieve your goals. It is important to your health and psychological well-being. Developing a healthy level of self-confidence can help you become successful in your personal and professional life.

Assertiveness

Assertiveness is the ability to confidently express your opinions positively, the quality of being self-assured and confident without being aggressive. *Dorland's Medical Dictionary* defines assertiveness as:

> a form of behaviour characterized by a confident declaration or affirmation of a statement without need of proof; this affirms the person's rights or point of view without either aggressively threatening the rights of another (assuming a position of dominance) or submissively permitting another to ignore or deny one's rights or point of view.

Assertiveness is a key communication skill that can help you to better manage yourself, people and situations. Being assertive is about being able to express yourself – your thoughts and feelings – effectively and stand up for your interests or point of view, while also respecting the rights and beliefs of others in a calm, direct and positive way, without being either overly aggressive or passively accepting 'wrong'. Simply put, it is respectfully and tactfully representing yourself – your opinions and recommendations – fully to others. It can help you to influence others to gain acceptance, agreement or behaviour change. You do not have to be aggressive (dominating or dismissing others to get what you want) nor be passive (failing to express yourself adequately, lacking self-confidence or timid). Assertive people are in control of themselves and are honest with themselves and others. They build strong relationships with others and allow others to feel heard and understood, even though they may not necessarily agree with them.

Assertive behaviour includes:

- Being free and open in expressing desires, thoughts and feelings and encouraging others to do likewise.
- Listening to the views of others and responding appropriately, whether in agreement with those views or not.
- Being able to respect the opinions of others as well as validate their feelings.
- Accepting responsibilities and being able to delegate to others.
- Regularly expressing appreciation of others for what they have done or are doing.
- Ability to admit to mistakes and apologize.
- Maintaining self-control.
- Problem-solving and compromise (compromise shows that the other person's needs have been heard and this is the attempt at providing a solution that all can be content with).

9.3 Benefits of Being Assertive

Assertiveness improves your communication and decision-making skills, helps boost your confidence levels naturally and keeps people from taking advantage of you in your organization. There is a direct relationship between assertiveness and self-confidence. It helps to eliminate or reduce the probability of being a trouble-maker within your organization and puts you in check so you do not act like a bully to others.

9.4 Developing Self-Confidence

Most women in the field of science and education are likely to feel confident:

(i) In their research area or area of expertise especially when they have good knowledge of what they are handling or what they are talking about. Knowledge is key and power, and there is the need to have a mastery of your field of endeavour. A well-informed mind cannot be deformed and can challenge a contrary opinion with available facts and figures. The rule of 'do it yourself (DIY)' should be observed at all times in the world of academia as this is paramount to developing self-confidence. When you are the team leader in your research or involved from the beginning to the end with a sound knowledge of the aim, objectives and data analysis, you will be enthusiastic and confident to disseminate information concerning that research. The same effect will be produced when you are involved in the packaging and editing of your presentations.

(ii) When they are excelling in an area or do something they have previously done very well. The better you are at doing something, the more confident you become. Success can induce a state of euphoria and is a propelling force to greater achievements. Excellent delivery of jobs or services in record time in the past will boost self-confidence. As you complete tasks and goals, your confidence that you can complete the same and similar tasks again increases.

(iii) When they are within the circle of people they know, trust and relate with freely. Self-confidence comes with the knowledge of acceptance irrespective of past mistakes and failures. When you surround yourself with positive-minded or people who exude positive energy, the possibility of excelling is very high as they will critic constructively to bring out the best in you.

There are odd chances out there against women, and you all have to learn to take responsibility for your careers. You must present yourself at any given opportunity – be it promotion, job hunt, change of job or career – and not feel intimidated by the number of your male colleagues with similar interests. You need to build your self-confidence and realize that you can do and achieve whatever you want. You must not give up when there are stumbling blocks or clogs in your wheel of our progress.

Sometimes it is difficult to figure out exactly where the lack of confidence comes from. It could be any or a combination of the following:

- Lack of information
- Low self-esteem
- Lack of preparedness

- Past failure experience and dented image
- The thought of taking up a task or making a change
- Being faced with an uphill task or a new challenge
- ▌ ▌▌▌▌ ▌▌▌▌▌ ▌▌ ▌▌ ▌▌▌▌▌▌▌▌▌ ▌▌▌ ▌▌▌▌▌▌▌ ▌▌▌▌▌▌▌▌▌▌▌▌ ▌▌ ▌▌▌▌▌▌
- When you are to make a presentation to your superiors
- When your integrity is questioned

These could fill you with so much fear and anxiety, making you shrink back into your comfort zone. However, there are ways of dealing with these situations:

1. Recognize and challenge your beliefs about yourself.

The biggest barrier to self-confidence is the **'I-Factor'**. The belief 'I' am not a confident person, 'I' am certain I cannot do this, 'I' will not be able to do this, 'I' am afraid, or 'I' am not sure I can do this. Confront your fears and ask what you are so afraid of. When you break it down, it may be something you can deal with. However, you have to step out of your comfort zone to try to succeed. You have to challenge this negative 'I-Factor' threat. Positive thoughts about yourself and the people around you can be a very powerful way of improving confidence. Try to highlight your strengths and successes and learn from your weaknesses and mistakes. Do not dwell on your failures and wallow in misery and regrets; dwell often on things that make you happy from your past while looking forward to more achievements. Negative thoughts about yourself can be very damaging to confidence and your ability to achieve goals. If you believe that you can plan and execute a project, you are likely to work hard to make sure you achieve it. However, if you lack the courage and do not believe that you can accomplish that task, then you are more likely to approach it with little or no seriousness and that marks the beginning of failure. You have to convince yourself that you can take up a task and accomplish it given the requisite tools and your state of preparedness and knowledge. It is often said that optimism is the faith that leads to achievement. Training, learning, practice and research can boost your confidence and help you to feel more confident about your ability to handle situations, roles and tasks. Therefore, challenge this negative 'I-Factor' threat and overcome it by thinking positively and declaring yourself a success.

2. Recognize your ability, talents and successes.

You are probably going to lose your self-confidence over time if you do not practice your skills and talents or if you hit setbacks. Constant practice of your skills and talents will help to maintain and boost your confidence further. It is important to know where your strength lies. Focus on and highlight your strengths. This helps you achieve more personal satisfaction and helps others build their confidence in you. Avoid telling yourself about your weaknesses all the time, and do not dwell on them; otherwise, you will lose the ability to recognize your strengths and successes, however small. Potentially or new difficult situations, as well as planning and preparing for the unknown, are perhaps the most important factors in developing self-confidence. Knowledge of what to expect and the processes involved in executing a task will make you feel more prepared and ultimately more confident. Although acquiring knowledge can sometimes make you feel less confident about your abilities to perform roles and tasks, when this happens, you need to combine your knowledge with experience and sometimes seek counsel from your mentors. By doing something we have learned a lot about, we put theory to practice which develops confidence and adds to the learning and comprehension.

3. Accept small challenges at a time.

Take some small risks to gain huge rewards. Instead of focusing on how you will make a mistake and not get it right, focus on what you stand to gain when you succeed. You do not have to be classified as an extrovert before you engage in advocacy projects. For many women, speaking to a group of people outside their clique or addressing a male-dominated audience is a particularly scary prospect, and the tendency to lose confidence is quite high. The best way to overcome this fear and gain confidence is with constant practice and experience. Face your fears and take action. Tackling your challenges usually includes facing one or more fears. You need to push or project yourself little by little, as you progress in small steps, this will sometimes increase your confidence, very considerably. For example, accept to anchor your group event or team meeting; volunteer to organize a seminar, workshop or even a conference. Volunteer to give a presentation or make a speech during events; when you accept a task, it makes overcoming the underlying challenges easier. Believe in yourself and be more willing to try new things, choose your task or challenge, practice resilience and have a mentor or role model you can draw strength from. However, do not quit because quitters never win and winners never quit. Always avail yourself of the opportunity of taking up or creating a new task or challenge.

4. Prioritize your tasks

Prioritization is key in achieving long-term goals. In setting your priorities, you decide deliberately to do what is the most important task even when everything on your list feels crucial. According to Brian Tracy, a productivity consultant, your monthly list pulls from your master list. Your weekly list pulls from your monthly list, while your daily list pulls from your weekly list. This way, your daily priorities are always aligned with your bigger goals. This prioritization method also helps combat the completion bias which is your tendency to focus on finishing small tasks rather than working on larger, more complex ones. When your daily tasks are being pulled from a larger list, you are certain of always working on meaningful and not just urgent things. You know what tasks need to be done by ranking them or listing them on a scale of preference. This will help you optimize your time, complete the most important and the most urgent tasks first and focus on the work that matters most. The chances of you becoming side-tracked when focusing on your priority list are very slim, and you will gain more of a sense of purpose. Learning to make informed decisions and being well organized and focused on accomplishing a task can boost your self-confidence. You can start or revisit a task or project that you have put on hold; this may seem overwhelming, difficult or awkward to complete, but simply making a start on such a task can boost your confidence and make you more inclined to accomplish it.

Thoughtful prioritization typically involves creating an agenda, evaluating tasks and allocating time and work to foing the most value in a short amount of time. You may adopt some of these strategies for prioritizing your tasks:

1. Make a to-do list that contains both personal and workday tasks in a single task list.
2. Identify the most pertinent tasks.
3. Identify which tasks must be completed promptly and plan according to future deadlines.
4. Prioritize based on importance and urgency.
5. Avoid dual-task strategy or competing priorities.
6. Evaluate tasks according to the effort required to complete them.
7. Do a critical reflection; frequently review your task list and priorities.
8. Prioritize your time and be realistic.

It is very pertinent to prioritize your tasks as this will help reduce stress and boost your confidence in terms of adequate preparations and meeting deadlines.

5. Communication

Communication is critically important and should not be taken for granted; it is the most essential tool in research and career development and advancement. Unfortunately, most research findings end on the pages of journals and are purely for academic exercises; they are not properly disseminated due to a lack of self-confidence. Self-confidence can mean not feeling threatened or intimidated by the presence of anybody. It allows you to speak concisely and with clarity when making public enlightenment campaigns, press briefings, presentations during the conference(s) or granting of interviews. Effective communication, irrespective of the medium, is a skill that reveals and strengthens your personality. The ability to efficiently articulate and convey your thoughts or express your feelings in ways that your audience or listeners will understand is a function of your confidence level. Be aware of your visual, verbal and non-verbal mode of communication or language. In terms of non-verbal mode of communication or body language, stay relaxed, stand upright and try not to wrap your arms around your body or slouch when making a presentation. When you are speaking, take deep breaths and speak slowly, at a steady and comprehensible pace and purposefully. Practice breathing rhythmically and in time with your speech.

6. Appearance

Confidence comes from within, but your appearance does have a profound impact on your level of confidence, assertiveness and overall behaviour. The wrong perception about your appearance can lead to great distress or low self-esteem. Personal appearance is an important factor in building your self-confidence. People who lack confidence tend to hide and do not want to be noticed. Learn to make a statement with your appearance and stand out in a crowd. When you dress and look good, you automatically feel better and proud of yourself both on the inside and on the outside. This has an overall boost on your confidence level, self-image as well as attitude towards people around you. Generally, your dressing reflects the way you feel about yourself or your mood. Dress moderately, comfortably and appropriately to suit the occasion as this can affect your mood and performance as well as the way others perceive/interact with you. The way you dress will determine how you will be addressed. It will be inappropriate to make a scientific presentation on the use of laboratory equipment wearing a dinner gown, a fascinator and a six-inch high heel shoe. On the other hand, wearing laboratory attire or an undersized or oversized dress, unkempt hair and a house shoe to boardroom presentation or dinner organized by your organization will attract negative comments. Consider which of these you could develop and vow to do something about it today: your appearance (clothes, hairstyle, gait and posture), the way you speak (tone, how you project your voice, words), your energy and enthusiasm and how expert or knowledgeable you are at something.

9.5   Developing Assertiveness

Impressions can make people think that you are confident. This can be achieved by assertiveness and can be applied to any situation where communication is key, for example:

• Meetings
• Presentations
• Interviews
• Dealing with colleagues
• Running projects
• Working with others

When you are assertive, you are self-assured and draw power from this to get your point across firmly, fairly and with empathy. When you have an opportunity to head a faculty board or lead an organization, be sure to do it well; be prepared to comfortably and honestly express your feelings. It is not about being aggressive or passively accepting wrong; it is about learning to be heard by consistently being a valued contributor and also express personal rights without denying the rights of others. Assertiveness can be learned, and several techniques can help you to develop your assertiveness:

1. Be your change agent.

Deciding to change your behaviour is a strong stimulator of change in itself. As a starting point, focus on one behaviour characteristic you would like to change and practice in a safe environment. The woman in academia can decide no longer to resigned to the situation and passively accept the skewed appointments of male faculty members into high profile positions but be a serious advocate against gender

bias and at the same time promote gender equality in terms of appointments into very visible positions in the organization. Sometimes your perception and approach to issues always result in a negative outcome. This can only change when you delib-
ⁱⁱⁱⁱⁱⁱⁱⁱ ⁱⁱⁱ ⁱⁱⁱ ⁱⁱⁱ ⁱⁱⁱ ⁱⁱⁱⁱⁱ ⁱⁱⁱ ⁱⁱⁱⁱⁱⁱⁱⁱ ⁱⁱⁱ ⁱⁱⁱⁱⁱⁱ ⁱⁱⁱⁱⁱⁱ ⁱⁱⁱⁱⁱ
resolve to change your approach. You will get the same result if you do the same thing the same way all the time. However, your result will change when you vary your approach and methodology. The ability to discover and accept your deficiency is the key agent you need for effective transformation; therefore, be your change agent.

2. Develop your confidence.

Positive thinking, or having an optimistic attitude, is the practice of focusing on the good in any given situation. It is an emotional and mental attitude that yields results that will benefit you and can also have a big impact on your physical well-being. It is essentially training yourself to adopt an abundance mindset and cultivate gratitude for your successes and those of others without you ignoring reality or making light of associated problems. Positive thinking can make or break an individual. Your thoughts affect your actions. Your actions, in turn, translate into whether or not you succeed in your field as well as influence the quality of your relationships and how you view the world at large. Feeling confident starts from within with positive thinking. When you are building confidence, learn how to reframe your mindset to start thinking more positively and feel more self-assured as a result. Think positively, spend time with positive people, focus on the good things, and practice gratitude. Challenging situations and obstacles are a part of life. Identify your areas of negativity, work on changing those negative thoughts you have about yourself to positive self-statements, practice positive self-talk, and start every day on a positive note. For example, instead of thinking 'I' am no good at presentations', tell yourself 'I can do this. You must think something positive about yourself. You can begin by focusing on all the reasons you will succeed instead of why you will fail. The more you repeat positive self-talk, the more you will believe it. Once you build this habit, you will be able to view failures and setbacks as stepping stones to success. This positive attitude can have a surprising impact on your confidence and help you to develop your assertiveness.

3. Control your emotions

Emotions can be positively or negatively expressed. Some people are very emotional while others are not; you must understand your emotions and those of others around you. When you understand that your emotions impact your behaviour, it can give you more control over them. Assertive people are not afraid to defend their points of view and goals or encourage others to see their point of view; they are also mindful of the feelings of others by watching their non-verbal communication mode which can be seen from their body language and facial expression. Assertiveness is about controlling your emotions and expressing them appropriately. For example, knowing that you are quick to rise to the bait in confrontational situations can help you develop strategies to cope with such or similar situations when they arise. Being assertive can help you control your emotions especially anger and stress and improve your ability and skills to cope with challenging situations. Memories, values or experiences might be behind some of your emotional outbursts. Take time to evaluate these factors as they affect your emotions and deal with each of them accordingly.

4. Communicate assertively, effectively and audibly

People who speak assertively, effectively and audibly send the message that they have self-confidence and cannot be intimidated and they are not overbearing or will stampede people to do their bidding. They know that their feelings and ideas matter. You must use plain language for effective communication. Do not use overly complicated words or acronyms that the receiver may not understand. The essence is to get your message across effectively. Use several simple but effective communication techniques or assertive statements such as 'I think', 'In my opinion', 'I understand' and 'I feel'. Think about what you are going to say before and how best you can say it for it to be effective, and this you must do constructively and sensitively irrespective of the content of your message. You do not have to be afraid to stand up for yourself and to confront people who challenge your opinion or your rights. You may get angry but be sure to control your emotions and to stay respectful at all times. Be clear, distinct, direct and accurate and always keep your requests simple and specific. Practice speaking out loud. Try this a few times to improve any point you have picked up on. You can write out some 'I' statements and empathic statements, or try speaking to your mentor, a group of friends or in front of a mirror to rehearse getting your tone right – polite, yet firm. Make a video recording of you speaking your argument out loud. Sometimes it is difficult to say no or turn down requests; you can try saying, 'No, I can't do that now'. Do not hesitate – be direct. If an explanation is appropriate, keep it brief.

5. Welcome others' opinions

Equal communication, negotiation and compromise are fundamental to assertiveness. It is important to be courteous to the other person and not talk in a frustrated, sarcastic or patronizing way. It is not necessary to constantly state your opinion or to assert your views in every situation. Communicating assertively involves putting your view across clearly while also taking into account the other person's perspective. There should be freedom of expression when relating with others irrespective of their position and opinion on a particular subject. You can use assertive techniques to say 'no', to give an alternative view or to point out things you feel are not right or to express positive thoughts and feelings. Be sure to use empathic statements such as 'I can understand how you feel', 'I appreciate your thought and contribution' or 'I want to plead with you'. You are only being assertive if you stand up for your rights in a way that does not violate the rights of others. Practice what you are going to say and how you are going to say it. If you consider it difficult to say what you want, you can practice general scenarios you encounter. Say what you want to say out loud. You can write it out, practice from a script or get a confidant to read and edit before going public. Consider role-playing with a friend or colleague and ask for clear feedback.

6. Evaluate your progress

Learn to evaluate your progress every time you try out your assertiveness; you may ask a confidant to also help in evaluating your progress. Questions you should ask yourself [?] 'how did I fare during this interaction?', 'how did I handle that?', 'What did I do well?' and 'what might I do differently next time?' Be open to criticism and compliments; accept both positive and negative feedback graciously, humbly and positively. Sometimes criticisms are inevitable and can draw out some emotional reactions. If you do not agree with the criticism that you receive, then you need to have a counter or superior argument, but without getting defensive or angry. You can always use feedback to keep you on track and help you identify areas for development and also to achieve significant positive change.

7. Accept that failures or setbacks are inevitable

Learn from the onset to accept that failures or setbacks are inevitable as you try out your assertiveness. It is difficult not to get frustrated or upset when you fail or hit a brick. Allow yourself to feel those emotions; however, do not let them get you down, but learn from them, and see them as stepping stones in your journey and that they are part of your developmental processes. Dwelling on your perceived failures and setbacks only makes you more depressed. There is no point beating yourself up when you should be figuring out how to keep moving. You need to be very honest with yourself. Review the processes or events that led to your failure and ask yourself what role did you play or where did you go wrong and why. Take complete ownership of the situation, and avoid repeating the same mistake. It is very important to stop blaming yourself or recapping your mess; forgive yourself and move on before you fall into that deep pit of hopeless despair. Let your setback be a motivator for you to double down and push harder especially when you are aiming for long-term achievement. Setbacks and failures can also boost our confidence; they make you realize that your other successes were not due to luck, coincidence or chance, but by you achieving a high level of accomplishments in your particular field. Setbacks should not be wasted opportunities; rather, they should motivate you to achieve something greater than what you had originally planned. Remember, failure does not mean it is all over; it simply means that it is time to re-strategize and do things differently. Be focused and resolute to overcome your failures and setbacks. It is also very important to recognize your successes and keep your failures in perspective because it is a learning and growing opportunity that is necessary for growth.

Robert Spadinger at Pick the Brain has a list of truths that can help adjust your definition of failure:

- Failure is an integral part of the way to success and self-realization.
- Whenever you step outside the comfort zone and whenever you try something new, failure becomes inevitable.
- Each failure brings you one step closer to reaching your goals.
- Failure is a great teacher and it allows you to learn some of the most valuable life lessons.
- Each failure makes you stronger, bigger and better.

- Making mistakes is not a big deal as long as you learn from them and avoid repeating them.
- Failure teaches you that a certain approach may not be ideal for a specific situation and that there are better approaches.
- Successful people will never laugh at you or judge you when you fail because they have already been there and they know about the valuable lessons you can learn from failure.
- No matter how often you fail, you are not a failure as long as you do not give up.
- Each time you fail, your fear of failure becomes smaller, which allows you to take on even bigger challenges.
- Every mistake is a learning opportunity, and after you have moved past your emotions, it is important to revisit your mistakes with a new perspective. Look at what you did that went wrong, but also look at what you did that was right, and what you can do better next time.

9.6 A Note of Caution

It is good to develop self-confidence because it is an important aspect of assertiveness. However, do not become too assertive; this may lead to arrogance or develop into a sense of self-importance; you may think you have the superior opinion and begin to stop listening to others despite them having good ideas. This will only act to alienate your colleagues and damage relationships. Your entitlements, opinions, thoughts, feelings, needs and wants are as important to you, but not more important than anyone else's. A little assertion at the right time can be a highly effective way of developing your profile and self-confidence.

References

Albion, M. W. (2008). *The Florida Life of Thomas Edison*. Gainesville: University Press of Florida. ISBN 978-0-8130-3259-7.

Assertiveness - An Introduction. (2011). Available from: https://www.skillsyouneed.com/ps/assertiveness.html [22 June 2020].

Axelrod, J. (2010). Building Assertiveness in 4 Steps. Available from: https://psychcentral.com/blog/building-assertiveness-in-4-steps/ [2 February 2021].

Black, C. & Islam, A. (2014). *Women in academia: what does it take to reach the top*? The Guardian International Edition. Available from: https://www.theguardian.com/higher-education-network/blog/2014/feb/24/women-academia-promotion-cambridge. [6 February 2021].

Brain, T. (2017). The Power of Positive Thinking: How Thoughts Can Change Your Life. Available from: https://www.briantracy.com/blog/personal-success/positive-attitude-happy-people-positive-thinking/ [9 January 2021].

Brain, T. (2020). How to prioritize tasks with a To-Do List. Available from: https://www.briantracy.com/ [9 January 2021].

Building Confidence. (2011). Available from: https://www.skillsyouneed.com/ps/confidence.html [22 June 2020].

Deep, P. (2016). 6 Proven strategies to rebound from failure. Available from: https://www.entrepreneur.com/article/285244 [23 March 2021].

Easy ways to build self-confidence. (2018). Available from: https://www.liveyourtruestory.com/11-easy-ways-to-build-self-confidence/ [14 June, 2020].

Frances, B. (2019). 5 Ways to be a more positive person. Available from: https://www.forbes.com/sites/francesbridges/2019/03/29/5-ways-to-be-a-more-positive-person/?sh=30feb27a716d [30 January 2021].

How to be assertive: asking for what you want firmly and fairly. (2018). Available from: https://www.cnblogs.com/kungfupanda/p/9166539.html [5 March 2021].

Jones, N. P., Papadakis, A. A., Orr, C. A. & Strauman, T. J. (2013). Cognitive Processes in Response to Goal Failure: A Study of Ruminative Thought and its Affective Consequences. *Journal of social and clinical psychology.* 32(5):10.1521/jscp.2013.32.5.482. doi:10.1521/jscp.2013.32.5.482.

Krause, M. (2019). Setbacks are necessary for success. Available from: https://medium.com/swlh/setbacks-are-necessary-for-success-3d845f511ec0%20%5b12; https://medium.com/swlh/setbacks-are-necessary-for-success-3d845f511ec0 [12 February 2021].

Mayo Clinic Staff. (2020). Being assertive: Reduce stress, communicate better. Available from: https://www.mayoclinic.org/assertive/art-20044644/in-depth/art-20044644#:~:text=Being%20assertive%20is%20a%20core,esteem%20and%20earn%20others'%20respect. [20 January 2021].

Mind Tools. (2020). Building self-confidence: Preparing yourself for success! Available from: http://www.mindtools.com/selfconf.html [5 March, 2021].

Robins, T. (2016). Positive thinking. 5 Strategies for positive thinking. Available from: https://www.tonyrobbins.com/positive-thinking/ [10 January, 2021].

Robins, T. (2020). The ultimate guide to building confidence: How to build confidence. Available from: https://www.tonyrobbins.com/building-confidence/ [5 June, 2020].

Sandberg, S. (2013). *Lean in: Women, work, and the will to lead* (1st ed.). Alfred A. Knopf.

Santos-Longhurst, A. (2019). Benefits of thinking positively, and how to do It: Medically reviewed by Timothy J. Legg. Available from: https://www.healthline.com/health/how-to-think-positive#overview [30 January 2021].

Spadinger, R. (2017). Available from: https://motema-safaris.com/2017/10/28/overcoming-your-failure/ [13 February 2021].

Tucker-Ladd, C. (2015). Building Assertiveness in 4 Steps. Available from: http://bmindful.com/forum/thread/5025/building-assertiveness-4-steps%20%20%20%20%20%5b15; http://bmindful.com/forum/thread/5025/building-assertiveness-4-steps [15 January 2021].

What is Self-Confidence? Available from: https://www.usf.edu/student-affairs/counseling-center/top-concerns/what-is-self-confidence.aspx [15 June 2020].

Chapter 10
The Secret to Being an Influencer as a Science Leader

Rachel Paterson and Uchechi Bliss Onyedikachi

10.1 Introduction

Different people have different views about who an influencer is. Some would say that an influencer is anyone on any of the social media platforms (Facebook, YouTube, Snapchat, Instagram, etc.) who has the ability to attract/persuade thousands and millions of followers to buy products and facilities (Peterson, 2020); others suggest that they are persons with referent power and ability to attract individuals and build trust (Latham, 2014). They have also been described as persons with the ability to affect the decisions of others due to their position, knowledge, and relationship with their target audience. Most times these individuals are solely assets in marketing and are therefore important in growing a brand. Influencers can be described by the number of followers, type of content, and level of influence/fame (Benchmark Report, 2021). An understanding of the word "influence" which entails the act of persuading and convincing people into having a particular thought is an essential part of leadership. Leadership is the ability to influence the behavior of others with or without resistance using different approaches to persuade their actions (Latham, 2014; Bacastow, 2018). A leader can use his/her power to affect people negatively or positively. This can benefit others or constrain them to work toward achieving organizational goals or to undermine them. A leader influences subordinates (downward power) and also tries to influence their boss (upward power). A science leader is not different from the leader described above except for his/her unique area of study that builds and organizes knowledge in the form of test,

R. Paterson
BetterManager, Professional Training & Coaching, San Francisco, CA, USA

U. B. Onyedikachi (✉)
Department of Biochemistry, College of Natural Sciences, Michael Okpara University of Agriculture, Umudike, Umuahia, Abia State, Nigeria
e-mail: ub.onyedikachi@mouau.edu.ng

© The Author(s), under exclusive license to Springer Nature Switzerland AG 2022
E. O. Nwaichi (ed.), *Science by Women*, Women in Engineering and Science, https://doi.org/10.1007/978-3-030-83032-8_10

description, principle, and predictions about the universe following systemic methodology based on evidence (Harper, 2014). Team play is necessary in successful leadership. A leader sets rules to make work done, while on the other hand subordinates (followers) embrace the vision and values of the leader by seeking to imitate more experienced coworker or trusted supervisor to achieve set goals (Bothomley et al., 2014).

A science leader who wants to be an influencer must be strong-willed, persuasive with great oratory prowess. Carving a niche in the life and lifestyle of followers should be sought by a science leader as this makes it hard to oppose or suppress influential leadership. A science leader/influencer must also communicate passionately both in writing and speaking. However, there is a high tendency that he/she may face obstacles like time constraints, unfamiliarity with platforms (e.g., someone may have inspiring materials to publish on a blog or on a LinkedIn profile, but may not know how to get his/her thoughts expressed to the audience or the target population), impatience, uncertainty about what to say (you know your thoughts, but how do you explain them efficiently), lack of confidence, etc.

Although these categories of people described are not fully the definition of who an influencer is, they still portray values and traits that could contribute to the criterion for rating an influencer.

An influencer is akin to opinion leadership. It is a leadership by an active media user who subsequently interprets the meaning of the media message or content to lower media users. Most people form their opinions from the information they gather from such leaders; hence, a science leader can become an influencer by sharing scientific findings to many people who are likely to become his/her followers as long as he/she makes positive impact in their way of life and style. This can be achieved through publications, video lectures on social media, magazines, etc.

Opinion leadership is formed from two models:

1. The two-step flow of communication propounded by Paul Lazarsfeld et al. (1944) and Elihu Katz (1957). Most people build their opinions under the influence of opinion leaders who in turn are influenced by the mass media. Here, ideas flow from the mass media to opinion leaders and then back to the people.
2. The one-step flow of communication, also known as the hypodermic needle model or magic bullet theory, holds that people are directly influenced by the mass media. An opinion leader actually has his/her opinion about issues/media contents especially in their field of interest (Baran & Davis, 2020).

In being an opinion leader, there must be followers/supporters or lower end media users. The power and voice of an opinion leader comes from the platform of his/her followers. This is because dedicated followers/supporters further disseminate the leader's message to the other media users. A science leader in like manner should have supporters who will help to blow his/her trumpet, hence creating more awareness on innovative findings as well as implementation methodology. Although the science leader creates a platform for himself/herself just like the opinion leader, the influence of his/her content lies in the network itself (Kuwashima, 2018). Opinion leaders can be either monomorphic or polymorphic. A monomorphic opinion leader is a follower of many other opinion leaders. Here, the opinion leader has access to far more information on a particular area than an average consumer. This is because he/she gains more access to first-class information/knowledge than others. However, a judgmental opinion leader is a leader. An unique opinion leaders. He/she can influence others in many areas (Flynn et al., 1996).

Factors of leadership includes:

(a) Expression of values
(b) Professional competence
(c) Nature of the social network

Opinion leadership seeks to make followers trust in order to interact and gain recognition from a social concept, thus improving his/her social status (Tushman, 1977). This is because people tend to possess status and sense of honor because of specific groups which they belong with distinct lifestyles and privileges (Ridgeway, 2014; Waters & Dagmar, 2016).

A science leader should be a boundary spanner. Boundary spanners are the link between the organization network and other sources of information. All systems have transference across their boundaries, and this process is facilitated by the boundary spanning. They bring in new ideas and innovations to the system (Matous & Wang, 2019).

Boundary spanners are needed to move knowledge around the organization/institution in a process referred to as socialization. The knowledge of research findings is needed within our various organizations, institutions, and journals. A boundary spanner who is an opinion leader/science leader is needed to transport such knowledge to the populace. This is known as internal boundary spanning, while in external boundary spanning, innovation systems are directly related to that of the organization. This depends on the absorptive capacity which can be defined as the firm's ability to recognize the value of new information, assimilate it, and apply it to its end users or commercial end (Cohen & Levinthal, 1990).

Absorptive capacity is one determining capacity of the degree of influence a science leader can make on the masses both on social media and on site. It is a strong predictor of innovation and knowledge transfer/infrastructure as well as the management support (Zahra & George, 2002).

A science leader must ensure that novel knowledge is disseminated as managers/people/supporters find it difficult to assimilate knowledge/information that is irrelevant to the current demand. The idea may be new, but if it does not intervene or seek to solve problems ravaging the globe, it may not get significant visibility and as such becomes less impactful (Zou et al., 2018). For example, during the lockdown as a result of the coronavirus-induced pandemic which affected the whole world, a science leader who aspires to be influential can contribute to knowledge which will enable the fast eradication of this wild virus. The adsorptive capacity would be glaring; its influence on supporters would be laudable, thus making the science leader more influential.

There are various criteria in rating an influencer. These include:

10.2   It Is Important to Know Your Values

I) Feel endowed by the knowledge and skills you possess no matter the challenges

The greatest gift to having an influence is the personal feeling of your own significance and that of others. Maya Angelou once said, "You may encounter many defeats, but you must not be defeated. In fact, it may be necessary to encounter the defeats, so you can know who you are, what you can rise from, how you can still come out of it." On your journey toward success, you are going to encounter some hiccups and missteps. Failing many times before you reach your end goal is far more common than you think. The trick is to stick with what you are working at, endlessly pushing. When people talk about the great innovators and inventors from history, they usually talk about their success. Their success started from knowing what they want to achieve and their values. So, before you intend becoming an influencer, know yourself, know what you intend achieving, and lastly know your values. According to an author of a book, *The Girl Who Found Water*, Azubuike, (2015), you cannot give what you do not have. It is easier to impact others based on a person's wealth of experience and knowledge. If you do not possess significant knowledge and skills which can be imparted into others, you might be unable to make potent effect on them.

II) Believe in yourself

Also, you must believe in yourself, love and appreciate what you have, and envisage positive impact on your target populace. This makes you bold, fluent, articulate, and strong-willed. If this is assimilated by a science leader, he/she would become influential. Even more so, a science leader who possesses this trait will become able to convince his/her target audience and as a result pulls many more followers. It is pertinent to know that our desire to make positive impact has the ability to empower us to do the needful. It is not just the desire to impact that announces your influence but also recognition from others via awards, certification, and media in both scientific field and outside.

III) Identifying core values

Every influential leader has core personal values. Core personal values are the basic belief of an individual or an organized group of people. These piloting principles are responsible for the character trait portrayed by a person (Anwar & Hansu, 2013). Parents also play a vital role on imparting positive core values to their children, thereby guiding them through living a good life.

The core values help people understand the difference between right and wrong. Core values usually guide people into determining the decision or choices they make toward realization of their goals. The examples of such core values are commitment, optimism, passion, honesty, efficiency, reliability, dependability, positivity, trust, loyalty, consistence, innovation, creativity, perseverance, respect, fitness, patriotism, peace, comfort, and belonging. There are also negative core values. This includes selfishness, greed, fear and insecurity, misconception about life, etc.

Another example of a negative core value is a tobacco company that supplies cigarette to the communities with regard to the effect and impact on health (Trach, 2015).

Some other core values pertaining to life are:

(a) A belief in God and connection to religious institution.
(b) A belief in applying frugality in our day-to-day life.
(c) A belief that family is important and vital to our livelihood.
(d) A belief that maintaining a healthy work and life is key to longevity.
(e) A belief that honesty is the best attitude to earning unwavering trust.

A leading scientist can also have cooperative core values (Tessema et al., 2019). One example is commitment. A commitment to sustain the environment by doing things that could contribute and improve research activities. For example, a scientist can sustain the environment by introducing the use of paper bags which are biodegradable rather than the use of nylon bags which are not biodegradable thus reducing environmental pollution. In addition, a commitment to doing well shows integrity, and this is paramount in leadership.

A science leader should be committed to innovation for excellence by carrying out research work with utmost care and dexterity. This will enable produce results that are factual and devoid of errors, thereby publishing research findings which can be trusted.

A science leader can show commitment by assisting indigent people by assisting with intellectual support that could improve their well being as well as the nation's economy. Also, a commitment to societal development especially in the educational sector can improve the quality of Scholars trained in various institutions of learning.

There are many core values, but it is alright to restrict to some in order to focus on our mission in life without distractions. However, working hard to increase positive core values can polish an individual to becoming the best he/she can be. The best way to identifying core values is by studying an individual's behavior (Nelson, 2019). To know what your core values are, ask yourself the following:

What activities bring you the most joy, or what you could not live without?
What gives your life a meaning, or what you want to achieve (Trach, 2015)?
What do I look out for in friendship?
What is most important in my work?
What are some of my favorite activities for the day?
What are my greatest failures, disappointments, and moments of depression and sadness?
What happened?
Was there some violation in some of your important beliefs?
What was violated (Nelson, 2019)?

10.3   A Science Leader Must Have Confidence

I) Build on your core values to boost your confidence.

If he/she has good knowledge of their core values, goals, and interest, their confidence is built. A science leader who bears this in mind and believes in himself/herself would be confident. The human mind is stirred up to do greatly in life's endeavors no matter how difficult only if one has the conviction that "you can." This becomes a boost in carrying out scientific research. Many successful scientists who made giant strides like Thomas Edison (Wooldridge, 2016) and Marie Curie (Marie Curie Facts, 2019) had challenging and difficulties before they rose to limelight (Peterson, 2020).

II) Be keen to learn.

A scientist who wants to be influential must be keen and open to learn from more informed colleagues in a particular field of interest. Science is vast, and as such, you cannot be a pro or an expert in every field. It is therefore advised that a science leader should be in the dark in solving scientific questions/research interest and findings. Someone may know or have thought deeper concerning your anticipated research interest without your knowledge (Park, 2003). Therefore, there is wisdom in sharing. There is always solution to every problem. Remember no man is an island. A science leader is sure to have a better result and outcome if he/she gets others involved in solving scientific problems (Varner & Peck, 2003). It also pronounces the visibility of the leader in the scientific world, even among colleagues. This potential makes a science leader a voice resulting in making salient contribution to the science not just that it opens one up to positive collaboration with other scientist further boosting his/her confidence, thereby leveraging diverse skills and strength (Miglianico et al., 2020).

Having confidence in the workplace or in career as a scientist can make one matured and keen to doing more. This can help lift a scientist in his/her field, thus conquering fear, challenges, and anxiety, thereby standing tall in science.

In attaining leadership levels in science, confidence attained often boosts the performance, skills, and development in one's field of practice. Confidence can also be ascribed as a mindset and can be felt by trusting in one's self. This goes beyond your knowledge as it further creates opportunity to improve science development.

III) Strategize on how to improve your skills.

(a) Professional development.

Professional development helps one in acquiring new skill or the already existing/acquired skill. For example, as a good scientist, you need good knowledge of statistics to be able to analyze your scientific findings, and that way, you may need to do a training/course on statistics to guarantee your proficiency in writing.

(b) Practice skills learned from trainings.

A science leader can improve in public speech by practicing speech presentations. This helps in he/she becoming influential in his/her field of endeavor as eloquence is developed as a result. There are diverse opportunities made available in conferences scientific meetings etc. that boost the confidence of the science leader.

(c) Dress for success.

Your dress sense attracts your type. If you dress professionally while interacting with colleagues, fellow scientist in your field as well as your senior/superior may help influence the confidence you desire. It is also very important to note that casual dressing for scientific conferences, board meetings, presentations, and other important events where formal business attire is required should be avoided. What you wear can influence your thought and bargaining prowess, hormone levels, and heart rates (Hutson & Rodrigues, 2015). Studies have shown that your dressing can affect one's performance mentally and physically. However, such findings on cognition are mostly from small studies in the laboratory that have not yet been replicated or investigated scientifically (Slepian et al., 2015). Formal dressing enhances higher action identification levels, greater category inclusiveness, and global processing advantage. Thus, one's cognition is broadly influenced by the clothing worn to events and how they are interpreted. Slepian et al. (2015) reported a research done to assess cognitive reasoning using undergraduates. It was concluded that those who dressed formally portrayed increased abstract thinking than those who dressed casually. This further buttressed creativity and long-term strategizing in those dressed formally. This work suggested a feeling of empowerment as a result of right and appropriateness in dressing. In another work done by the *Journal of Experimental Biology*, male subjects who dressed in suit and their counterpart who dressed in usual duds (casual) engaged in a game of negotiating with a partner. Those dressed in suits obtained profitable deals and exhibited higher testosterone levels than those dressed casually. People would always judge others by their appearance; as a science leader with influence, your clothing should match your expectation per time. A particular audience or scenario should be the determining factor of what dress should be worn. For instance, if scientist is faced with convincing decision-makers, a top-notch suit (formal) may be suitable, but he/she should exercise caution not to overdo by dressing to impress because this may broadcast wrong priorities and in turn likely to become a distraction. Thus, at this point, a person wearing a pair of pants and T-shirts may seemingly become more accepted.

A science leader can also use his/her clothing to highlight his/her findings or interest, while this can be used in some other careers to describe a brand (Amsen, 2020). This in turn connects you more to your colleagues and the public. A young/early career scientist can also ruin a relationship with mentors if his/her dressing becomes a turnoff. He/she should not also dress to "outshine" their mentors. Dressing to look like what you are not makes you look unserious earning you losses at the end. Therefore, every science leader who aspires to be influential should be keen on dressing neatly and modestly. However, it should be noted that dressing well and being a successful scientist are not mutually exclusive. There are several strong and excellent scientists who are good dress lovers, and this has not affected their scientific output. When in the laboratory, a scientist should have a laboratory coat, goggles, mortar board, wacky science tie, laboratory timer, pocket protector, gloves, pencil, and BSL-4 positive pressure isolator.

IV) Dare to overcome your fears by leaving your comfort zone.

One of the easiest ways to boost confidence is by stepping out of your comfort zone. A scientist who dreads to present papers at conferences and symposium, he/she can actually step out instead of shying away by attending conferences and also presenting papers too at every opportunity. He/she can also volunteer to speak on behalf of your research team in events and other scientific exhibitions. Attend seminars to improve on public speaking skills as this would help to tune your body language and improve your presentation delivery so that people can hear and understand so well. Stop using filler words like um, ah, ok, em, you know, etc. Do not also try to memorize scripts as this could be harmful if you speak a word; also, practice speaking slowly, and enunciate clearly and audibly. It helps one organize words, speech, and grammar. Speaking this way, one is unlikely to make mistakes (Cylon, 2021). Also, it is necessary to pause to ask question, and when you do that, wait to get an answer. This practice can help carry your audience along. In toning your body language, you can practice speaking without backing your audience, move slowly, and be relaxed; also, match your facial expression with the overall energy of what you are about to present (Genard, 2019).

Connect with the audience by making contact by looking at some persons while presenting. Use vocabulary and examples that are easy to understand. Also, a scientist can make his/her presentation interesting by using funny anecdote, stories, visual aids and characters are also encouraged. Did you know that every speaker has a degree of anxiety especially when presenting in an unfamiliar ground? Therefore, practicing many times is one important character of a good public speaker. A science leader who wants to influence people should therefore adopt the style of reading aloud over and over again (Raso, 2013). He/she can record and rehearse. This helps to conquer anxiety; it surely would boost his/her confidence.

V) Emulate confident peers.

An aspiring successful scientist can emulate strategies used by other senior science colleagues by observing their mannerisms and quickly applying it to his/her own career. For example, adopt several ways that successful colleagues comport/present themselves in sitting, standing, and talking and practice the same. It will influence your overall confidence as a scientist (Peterson, 2020).

VI) Set goals for yourself.

Evaluating small successes helps project one to do more. A science leader can target objectives by remodeling his/her time and schedule, thereby increasing productivity (Peterson, 2020).

VII) Focus on your strengths, not your weaknesses (Roberts et al., 2005).

VIII) Eliminate negative language.

Always applaud yourself for successes and achievements no matter how little. You can also change your thought by setting possible and feasible goals and do not be overly critical about yourself. Take your time. Be persistent, and keep developing your mindset.

IX) Always ask questions to relieve anxiety.

10.4 Maximize Your Potentials or Influential Opportunity

I) Identify the talents and strengths in others.

Great influencers have foresight of identifying the talents and strengths in others and quickly taking advantage of it to develop oneself. Miglianico et al., (2020) in their work on ascertaining strength stated that strength contributes to the development; their use does not diminish another person. They manifest at cognitive, emotional, and behavioral level. They are desired and cultivated by different cultures. A school of thought was developed at the center of applied positive psychology founded by Alex Linley (2008) to assess the strength of an individual. This focuses on energy, performance, and use which are vital in defining a strength used by different people to achieve success.

II) Identify your strength.

This approach defines strength as pre-existing capacity for a particular thing or way of behaving, thinking, or feeling that is authentic and energizing to the user and enables optimal functioning, development, and performance. A person's greatest potential lies in his or her strength. Therefore, researchers and science leaders are invited to refine and develop the actual knowledge in the field, while professionals are encouraged to identify use and optimize employee's strength in organizations through improved resource management practices (Miglianico et al., 2020).

In developing your strength as a scientist who wants to influence others in career, a tool to help you understand and leverage on your strength is called Reflected Best Self (RBS). This tool helps you tap into some of your hidden talents while increasing your career potential. The feeling of exhaustion and disappointment always demoralizes a person especially when faced with some scientific problems. Sometimes, from my experience, it is usually challenging getting a mentor who is interested in you as much as you are interested in them. In other scenarios, the mentor does not have enough time at their disposal to mentor a mentee. Some mentors also portray perfectionism by rather expecting so much from a younger colleague. This is rather frustrating, and this approach hampers the abilities and potentials of a growing scientist. Therefore, the RBS tool (Roberts et al., 2005) can be adopted to encourage a science leader gearing toward development, hence improving on their strength via the following steps:

(a) Identify respondents and ask for feedback.

This is done by collecting and analyzing inputs from persons like i.e., both inside and outside work about strength. The feedbacks gotten through this tool indicate that people usually have/possess more strength than they know (Roberts et al., 2005).

(b) Recognize the patterns.

This is done by analyzing the feedbacks, similar patterns and comments from your respondent. It is usually surprising when people tend to remember your actions than you do. In this step, a science leader can be exposed to information from respondents with regard to your strength which in turn becomes a positive energy to encourage him/her to live up to expectation or improve oneself. The RBS exercise sheds light on those skills and talents ignored or taken for granted (Roberts et al. 2005).

(c) Compose your personal portrait.

This includes writing a description of who one is from feedbacks and other information gotten from respondents. It should not be just cognitive and psychological knowledge but rather insightful picture that tells you grounds broken (success and great picture of what the future holds). A personal portrait gives a summary of one's best. This helps to point out reasons for past failures and a variety of problems to be resolved thereby making positive changes in life (Roberts et al., 2005).

(d) Redesign your job.

After pinpointing your strength, redesign your personal job description. There is now a conscious relationship between your portrait of your best and work expectation. In redesigning, time management and proper scheduling of one's time make efficient. Thus, opening and prompting a leading scientist to collaborations with people of same vision and goals. The team play helps to improve strength and reignite passion to meet set goals. Reaching set goals often brings light and joy and gears one into achieving more. The need for knowing your strength cannot be exhausted; hence, a science leader is advised to build on them, while dealing with weaknesses in turn gaining confidence to do achieve beyond ones possible best (Roberts et al., 2005).

III) Follow scientist with admirable traits.

It is also important for a science leader who wants to be an influencer to follow scientist with admirable traits to remain inspired. McCabe (2016) says that participants completed scales measuring willingness to celebrate the success of the other and also a motivation to self-improve and also bolster self-esteem.

IV) Be intentional in improving on others.

In maximizing potential, be intentional in making others reach their full potential and also work to develop your ability to see the bigger picture of those you lead and those you wish to do business with. A science leader, commands influence and earns admirations for providing opportunities that improves and magnify the potentials of his or her followers. Also, empower your subordinates by entrusting them with privilege task and allowing them to do it. In believing and trusting those who look up to you the rest of your words of encouragement on like "you are able" can set them on the valiantly. Teamwork builds trust and trust builds growth. The growth and development of people represents one of the calling of leadership, just like in the case of a science leader (Peterson, 2020).

10.5   Effective Communication

Communication is another major way of influencing people today. It is an exchange of meanings accordingly for organizations and humans as a social being (Cross, 2018). Communication involves the exchange of information between individuals through a system of signs, behavior, and symbols. It may be verbal or nonverbal. Communication accesses the mind and the thought of another. It is important for successful relationship in the workplace and may include body movements, facial expression, posture, and interpersonal distance. As a growing science leader, communication is needed for exchange of opinions, making plans, proposals, reaching agreements, and executing decisions (Christopher & Corina, 2020). Effective communication improves performance, and it is also linked with strategic planning. To produce consensus interactions, individuals commit to an issue until a large majority agrees on an option.

Communication is the greatest vehicle for influencing others and society with your science. Every interaction with another person determines how you are perceived, and every interaction is an opportunity to develop trust and exert a positive influence. Whether presenting one-on-one or to an audience of one hundred, conveying information to a team, or delivering a difficult message, communicating effectively is one of the most powerful tools for achieving your objectives and having an influence.

Skills to Effective Communication

I) Create good rapport with the people you wish to influence by creating an enabling environment. This is because setting new and cordial relationship makes communication easier.

II) Listen while communicating: If you listen to people, they feel valued and more likely to be persuaded in believing in you.

III) Ask intelligent question and interesting ones too: Using the right question techniques gives you the right answers you desire.

IV) Beware of body language: Mirror the other person's body language to create better rapport.

V) Sell the benefits of your argument to the other person and try to see your position from their perspective.

VI) Be relaxed: A relaxed and a natural demeanor can enhance effective communication with ease. One can fully express himself/herself when relaxed.

10.6   Build Connections

As employees, we spend a significant part of our time in the workplace. Sincerely, the state of our relationships with colleagues affects us greatly, in our ability to succeed professionally and in our emotional well-being (Peterson, 2020).

Naturally then, it is in our best interests to form healthy, effective relationships with colleagues. This is not always going to be a straightforward task. The workplace brings together a mixed kind of people, not of our choosing, with different values, cultures, expectations, age ranges, and personalities – all sources of potential conflict. By practicing and building these skills, we can make our working lives a great deal more pleasant and efficient.

Cultivating personal connections with colleagues allows them to get to each other. One of the reasons people do things for others is relationship. It is critical to have good rapport with your colleagues. This will not translate directly into influence, of course, but it makes it more likely that others will at least hear what you have to say (Krishnaswamy, 2016).

It is far easier to become intentional about creating a connection culture when the importance of connection is understood and colleagues are aware of disconnected subcultures within the working environs.

Consequently, what ways can we build a healthy relationship with colleagues?

I. Mark the traits that are important for achieving a connection culture.
II. Circle any items you marked that need to be strengthened.
III. Identify individuals in your organization who could help you close those gaps.
IV. Prioritize the list of actions that you identified.

Focus first on your top three actions, reaching out to the individuals you identified to see if they can help you close those gaps. Then work your way down the priority list to continue increasing connection among colleagues. Inasmuch as you have gotten to know yourself and your values regarding to what you intend achieving, try to cultivate personal relationship or connections with your colleagues and allow them to get to know you. This may not directly increase your influence but it makes it more likely that people will pay more attention to what you have to say. As the saying goes, communication is the greatest vehicle for influencing people. Every interaction with another person determines how you are perceived, and also every interaction is an opportunity to develop trust and exert a positive influence. Whether presenting one-on-one or to an audience of one hundred, conveying information to a team, or delivering a difficult message, communicating effectively is one of the most powerful tools for achieving your objectives and having an influence. You can influence policy with your science crafted as evidence (Martino et al., 2017).

10.7   Listen Before You Try to Persuade

It has been said that attention is oxygen for relationships. The best way to prime colleagues for backing you and your agenda is to make them feel heard. Start by giving them your undivided attention in one-on-one situations. Practice what is called "the discipline of focus." To do this, face the other person (in person or virtually), freeze in place, and listen. A big part of workplace resentment is people feeling disrespected and that their voices are not being heard. So, ask colleagues for their perspectives and advice and then *listen* to it. It is pertinent to listen to people to help them change (Itzchakov & Kluger, 2018). Listen carefully, observing facial expressions and body cues (Tamerius, 2019).

People need affirmation and recognition, so get in the habit of looking for ways to affirm and serve others. Do this by looking for task strengths and character strengths, which reflect the excellence of someone's work and the way someone goes about his/her work, respectively (Allen, 2020). For example, you might affirm a colleague by saying, "That was an outstanding project you carried out, the paper was written simply and easy to understand." You might affirm his/her character strengths by saying, "I appreciate the way you persevered to make our project happen. You showed wisdom and humility in seeking the ideas of others and applying the best ideas to the design of our new website. Very nicely done."

Paying tribute where it is due will always create a healthy respect with colleagues. Just make sure not to use praise as a means of manipulating people to do what you want. Also, when you constantly speak positively about others, then people will notice this and recognize you as a good colleague or leader. Good relationships help develop a confident workplace where the environment empowers you to deal comfortably with any potential stressful situations.

You can also build connections with people during negotiations if you adopt and maintain the right mindset. Thinking of the people you are working with as competitors leads to disconnection and distrust. Instead, think of them as holding knowledge that you need in order to identify a win-win solution. Negotiating requires probing, patience, and perseverance to understand other people's objectives, perceptions, and sensitivities (Derrick & Wooley, 2009).

10.8   Mind Your Body Language (and Your Tone)

Our brains are constantly assessing whether to trust others or not; we are hardwired to be asking the question, "Is this person a friend or foe?" Your body language is critical to conveying the right message. How you stand (tall, arms uncrossed), sit, lean forward, and the pitch of your voice (slightly lower demonstrates confidence, slightly higher demonstrates nerves) can send signals to the other person (Peterson, 2020).

When interacting with others, make sure your body language is connecting with them by being still, maintaining eye contact while looking away at times so as not to look too intense, relaxing your facial muscles, uncrossing your arms, and leaning slightly forward. Agreeing with others is connecting, but it has to be a genuine agreement – so when you agree, indicate your agreement with a positive head nod, occasionally using words such as "yes," "I agree," or "absolutely."

This is a powerful Technology, Entertainment and Design (TED) lecture about the power of our body language (Amy, 2013). When you enter a room, and it is appropriate given the context and number of people present, take time to greet or nonverbally acknowledge each individual present, even when you are not familiar with them people. Not personally acknowledging them, either at the start or end of the meeting creates an impression that you are indifferent.

When you meet people for the first time, ask questions to identify something unique about them. In doing this, it will make you more likely to remember the person.

10.9 Develop Expertise

Increase your influence at work by immersing yourself in your topic area but in a public manner. Think about who knows you in your research area. Regularly attend industry conferences, enroll in a webinar or specialized program to broaden your knowledge of supporting topics, or take on a leadership role in a relevant professional organization or association – visible and public signs that you are staying up to date and informed.

Make peer mentors available for easy learning and improvement in a specific area of competence or character, and select a mentor who is strong in the given area.

Ways we can develop expertise are:

I) Set goals for yourself.

Setting specific goals to improve your career helps you stay on track with your development. Make sure your goals are measurable, achievable, and relevant to your profession or your goals. Then, consider organizing a timeline to achieve your goal by setting a beginning and end date, as well as smaller goals to achieve along the way (Roberts et al., 2005).

II) Find a mentor.

A professional mentor is typically a superior you respect and trust. Once you find your mentor, you can reach out for informal meetings, which can then naturally develop into a professional relationship.

III) Review job descriptions for positions you want.

These job descriptions will give you an idea of the transferable skills you have, as well as the job specific skills you will need. Once you identify the skills you need, you can research Job shadowing or education programs that can provide you with the necessary skill set to transition into that position.

IV) Study to improve yourself.

Companies often encourage employees to further their education with a degree, and some offer tuition assistance or reimbursement. If you are advancing your career with a related program, such as accounting and finance, you may also find that many of your credits are transferable.

V) Take continuing education courses in career-related fields.

These courses are often taught by professionals with experience in their field. For some professions, continuing education courses are required to stay current in the industry. Many colleges and universities offer continuing education courses in a variety of fields.

VI) Take advantage of company training.

Many companies use independent training departments with experts in different fields that train on specialized skill sets. Check in with your supervisor about what your company has to offer and which courses would be especially beneficial for your professional growth.

VII) Participate in job shadowing.

Job shadowing is a great way to learn more about the day-to-day responsibilities of another profession and to learn new skills. Generally, it is best to choose individuals who are experienced or perform well in their positions. Job shadowing usually involves following a professional while they perform their job duties and learning about different skills.

VIII) Join a professional association in your field.

In a group setting, you have the opportunity to converse with colleagues about your industry and to discover skills you may want to develop. These professional associations are usually available on local, state, national, and international levels.

10.10   Map a Strategy

When it comes time to leverage the influence you have built to promote a particular initiative or concept, be strategic. Create a plan or map to guide your "campaign," an organizational chart of the decision-makers related to your issue. Look at each level and ask yourself, "Can I influence this person directly?" If not, whom can I influence who can influence that person? Then strategize how and when you will approach these various colleagues. People are more engaged when they are striving and progressing toward goals. Help them make wise goals to advance their careers, and put disciplines in place to help them achieve these goals. Doing so will boost their effectiveness and connection to you (Knight, 2019).

I) Give people what they want.

Authentically frame your issue as a benefit to the people you want on your side. Consider each person's needs, perspectives, and temperaments. Align your research with your needs of the society. Find out what each person needs to hear and what will capture their attention. Remember they will each be thinking, "What's in it for me?" Use the word "we," as in "We'll see value."

If the individuals you are responsible for leading disagree with your decisions, seek them out and consider their opinions. This shows you value them and want to give them a voice. Considering their opinions may also better equip you to make better decisions and enhance productivity in the future. Whenever possible, include individuals who express interest in an issue (Knight, 2019).

II) Connect with people.

Connect authentically with other people. Cultivate personal connections with colleagues so they assume positive intent when you attempt to influence them. Make it clear to your colleagues, your collaborators, and members in your networks that you value their opinions.

III) Work hard.

Work hard to build trust and when you do guard it jealously be mindful of each statement made.

IV) Project your purpose.

Try not to force a concept on the people like the media is portrayed.

V) A leader must be firm.

The leader must have a more determining role in an individual's opinion or action. So be mindful not to be tossed around by external influential individual as this might even make an opinion leader loose his/her status.

10.11 Best Practices of a Successful Influencer

- Always consider the integrity and ethics of science, in order to influence others.
- Listen to people, and create a trusting and compassionate environment.
- Share a clear vision and knowledge, and are inclusive.
- Act as a positive role model for others.
- Give options and latitude, allowing for calculated risks.

- Take responsibility and ownership.
- Inspire and elevate your team members and the whole organization.
- Focus on solutions and avoid blaming others.
- Demonstrate resilience in facing adversity.
- Celebrate efforts, resilience, and success.
- Join a professional society and get involved.
- Develop your core leadership qualities.
- Develop your strengths.
- Seek opportunities to collaborate.
- Demonstrate your ability to innovate.
- Be proactive and not reactive.
- Find a mentor.
- Lead without authority.

10.12   Conclusion

The ability to lead and to influence others is no longer the prerogative of "the select few at the top." It is core capability, no matter what level you work at, and it is fundamental to purposefully engage with peers, superiors, reports, and decision-makers within and outside our own organizations. In fact, organizations are increasingly seeking this capability in the people they select and invest in. Today, maybe more so than ever before, anyone at any level with the right mindset, skills, and tools can influence and mobilize people and achieve greater meaningful and lasting impact even as a science leader.

References

Allen, T. (2020). *Persuasion: How to convince people to act on your great idea*. Forbes. Retrieved online 30th March 2021.

Amsen, E. (2020). Science and culture: Researchers embrace fashion to show off science concept. *Proceedings of the National Academy of Sciences of the United States of America (PNAS), 117*(13), 6959–6962.

Amy, C. (2013). *Does body language shape who you are?* TED Radio Hour. Retrieved 29 March 2021.

Anwar, J., & Hansu, S. (2013). Ideology, purpose, core values and leadership: How they influence the vision of an organization. *International Journal of Learning and Development, 3*, 3. ISSN 2164-4063.

Azubuike, C. M. (2015). *The girl who found water*. Memoirs of a Corps Member. ISBN:9789789439577.

Bacastow, C. (2018). *Boy Scouts of America: Training leaders for tomorrow*. Content and user contributions on this site are licensed under : CC BY: Attribution with attribution required.

Baran, S. J., & Davis, D. K. (2020). *Theories of mass communication. Introduction to mass communication*. McGraw-Hill Education. ISBN 978-1285052076-Via McGraw-Hill online learning centre.

Bothomley, K., Burgess, S. I., & Fox, M., III. (2014). Are the behaviours of transformational leaders impacting organization? A study of transformation leadership. *International Management Review, 10*(1), 519.

Christopher, K. T., & Corina, E. T. (2020). Social influence and interaction bias can drive emergent behavioural specialization and modular social networks across systems. *Journal of Royal Society Interface, 17*(162), 20190564.

Cohen and Levinthal. (1990). Absorptive capacity: A new perspective on learning and innovation. *Administrative Science Quarterly, 35*(1), 128–152.

Cross, O. D. (2018). Effects of communication strategies on the performance of the public Organizations in Nigeria. *International Journal of Science and Technology, 6*, 8. ISSN.2321-919X.

Cylon, G. (2021). *10 Ways to step out of your comfort zone and overcome your fear*. Success mindset. Last updated March 2, 2021

Derrick, S., & Wooley, K.. (2009). Meetings with Scott Derrick and Kitty Wooley of 13L.

Flynn, L. R., Goldsmith, R. E., & Eastman, J. K. (1996). Opinion leadership and opinion seekers: Two new measurement scales. *Journal of the Academy of Marketing Science, 2*, 147.

Genard, G. (2019). *The 5 key body language techniques of public speaking*. https://www.genard-method.com/blog/bid/144247/the-5-key-body-language-techniques-of-public-speaking. Retrieved 30 April 2021.

Harper, D. (2014). *"Science" online etymology dictionary*. Retrieved 29 march, 2021.

Hutson, Matthew & Rodriguez, Tori. (2015). Dress for Success. Scientific American Mind. 27. 13–13. https://doi.org/10.1038/scientificamericanin/0016-13a

Itzchakov, G, & Kluger A. N. (2018). *The power of listening in helping people change*. Harvard business review analytical services. Retrieved online 29th March, 2021.

Katz Elihu. (1957). *The two step flow of communication: An up-to date report on a hypothesis*. University of Pennsylvania Scholarly Commons.

Knight, K. (2019). *Sticky wicket (dress code)*. Florida Atlantic University, Boca Raton, FL 33434. Published by the company of Biologist Limited.

Knight, R. https://hbr.org/2018/02/how-to-increase-your-influence-at-work; https://blog.job-sgopublic.com/how-to-maintain-a-healthy-working-relationship-with-colleagues/Forbes. Beth review.

Krishnaswamy, A. (2016). Building connections. *Essay on Science and Society, 354*, 6312–6558. https://doi.org/10.1126/science.aa9763

Kuwashima, Y. (2018). The strength of an opinion leader's supporters. *Annals of Business Administrative Science, 17*(6), 241–250. https://doi.org/10.7880/abas.01810099

Latham, J. R. (2014). Leadership for quality and innovation: Challenges, theories and a framework for future research. *Quality Management Journal, 21*(1), 11–15.

Lazarsfed, P. F., Berelson, B., & Gaudet, H. (1944). *The people's choice: How the voter makes up his mind in a presidential campaign* (p. 151). Columbia University Press.

Linley, P. A. (2008). *Average to A+: Realising strengths in yourself and others*. CAPP Press.

Marie Curie Facts. Nobel prize.org. Archived from the original on 6 March 2019. Retrieved online 29 March 2021.

Martino, J., Pegg, J., & Frates, E. P. (2017). The connection prescription: Using the power of social interactions and the deep desire for connectedness to empower health and wellness. *American Journal of Lifestyle Medicine, 11*(6), 466–475.

Matous, P., & Wang, P. (2019). Extend exposure, boundary spanning and opinion leadership in remote communities: A network experiment. *Social Networks, 56*, 10–22.

McCabe, E. A. (2016). *What is the influence of admiration on the social comparison process?* Honors theses. 364.

Miglianico, M., Dubreu, I. P., Miquelon, P., Bakker, B. A., & Martin-Krumm, C. S. (2020). Strength use in the workplace: A literature review. *Journal of Happiness Studies, 21*, 737–764. https://doi.org/10.1007/s10902-019-00095-w.

Nelson, M. (2019). How to determine your core values. Your important lie assignment. *Noteworthy– The Journal Blog*. https://medium.com/@michellefyfe/how-to-determine-your-core-values-99f6392eea27

Park, C. (2003). Engaging students in learning process: The learning journal. *Journal of Geography in Higher Education, 27(3), 183–199.*

Peterson, R. (2020). *The secret to being an influencer as a science leader*. Organization for Women in Science for the Developing Nation, Nigeria, National chapter, University of Port-Harcourt Branch series of Scientific Communication. https://owsd.net/resources/news-events/owsd-nigeria-national-chapter-presents-secre-being-influencer-science-leader

Raso, R. (2013). Brace yourself! Leaving your comfort zone. *Nursing Management, 44*, 8-6. https://doi.org/10.1097/01.NUMA.0000432225.02005.5a. Retrieved online 30th March, 2021.

Ridgeway, C. (2014). Why status matters for inequality. *American Social Reviews, 79*, 1–16.

Roberts, M. L., Spreitzer, G., Dutton, E. J., Quinn, E. R., Heaphy, D. E., & Brianna, B. (2005). *How to play to strength*. Harvard Business Reviews.

Slepian, M. L., Ferber, N. S., Joshua, G., & Abraham, M. R. (2015). The cognitive consequences of formal clothing. *Social Psychological and Personality Science, 6*(6), 661–668. https://doi.org/10.1177/1948550615579462

Tamerius, K. (2019). *To persuade change the way you listen. Smart politics*, New York Times.

Tessema, M., Dhumal, P., Sauers, D., Tewolde, S., & Teckle, P. (2019). Analysis of corporate value statements: An empirical study. *International Journal of Corporate Governance, 10*, 2.

Trach, E. (2015). *What are core values of a company? Definition and examples*. http://study.com/academy/lesson. Retrieved 28 March 2021.

Tushman, M. L. (1977). Special boundary roles in the innovation process. *Administrative Science Quarterly, 22*(4), 587–605.

Varner, D., & Peck, S. R. (2003). Learning from learning Journals: The benefits and challenges of using learning assignments. *Journal of Management Education, 27*(1), 52–77.

Waters, T., & Dagmar, W. (2016). Are the terms "Social economic status" and social status warped form of reasoning for max weber? *Palgrave Communication, 2*, 16002.

Who is an influencer? Social media influencer (updated 2021). Influencer marketing hub. Benchmark report. https://influencermarketinghub.com.

Wooldridge, A. (2016). *The alphabet of success*. The Economist. Retrieved March 30, 2021.

Zahra and George. (2002). Absorptive capacity: A review, reconceptualization, and extension. *Academy of Management Review, 27*(2), 185–203.

Zou, T., Ertug, G., & George, G. (2018). The capacity to innovate: A meta-analysis of absorptive capacity. *Innovation, 20*(2), 87–121. https://doi.org/10.1080/14479338.2018.1428105.

Chapter 11
Ethics in Science Through the Lens of COVID 19 Pandemic

Florence Onyemachi Nduka

11.1 Introduction

Science has always been seen as pragmatic and the rampart in a surging ocean of confusing and polarizing ideas, and nothing has put science on the frontlines in recent times than the COVID-19 pandemic. Science and its practitioners have tried to live up to its goals of systematic observation, analyses, and logical conclusions. It depends on evidence and is not swayed by conflicts or political shades and persuasions. COVID-19 provides the perfect picture of the flux in scientific research and the painstaking efforts at observations, trying to understand the facts and proffering possible conclusions. COVID-19 from its inception in China in December 2019, becoming a global emergency in late January 2020, and finally upgraded to a pandemic on March 11, 2020, by the World Health Organization (WHO) has spurred the scientific world into action. Science provided the genetic sequence of the virus SARS-CoV-2 which led to a flurry in diagnostics, information on early detection and case managements were made available and updated as scenario changed, and issuance of public health guidelines and public safety measures. Though the pandemic came with unusual pressure and urgency, scientists knew they still have to operate within the ethics of the profession and uphold the moral guidelines that distinguish what is right and wrong. Major ethical principles operate within and across science disciplines and viewing through the COVID-19 lens, how have scientists fared in upholding these ethics as they race against time to limit the spread of this contagious virus with an elegant name? Science activities through the ethics pathway during this pandemic will be reviewed under the following headings (Weinbaum et al., 2019).

F. O. Nduka (✉)
University of Port Harcourt, Port Harcourt, Nigeria
e-mail: florence.nduka@uniport.edu.ng

E. O. Nwaichi (ed.), *Science by Women*, Women in Engineering and Science,
https://doi.org/10.1007/978-3-030-83032-8_11

11.2   Duty to Society

This ethical principle encourages researchers to embark on research that must contribute to the well-being of society. The race to stem the tide of COVID-19 showcases a duty to society. The concern of the majority, of the scientific researches embarked on was to stave the infection and improve the well-being of the affected. The genome of the virus was quickly analyzed to provide a road map towards understanding its nature. Science as a duty to society showed that the SARS-CoV-2 from phylogenetic relationships and structures of the genome belonged to the *Betacoronavirus* genera and had a single strand of positive sense RNA in a capsid comprised of matrix protein (Kamps & Hoffmann, 2021). This analyzing of genome pointed out the right pathway towards accurate diagnostic tests and long-term control.

The transmission pathway was quickly researched to understand how it moves from one host to another. Some hiccups were experienced given the novel nature of the virus, but serious effort was made at educating the public with the available information that it is a respiratory infection transmitted through droplets emitted through sneezing, coughing, and by touching contaminated surfaces. The symptoms of fever, chills, headache, coughing, sneezing, shortness of breath or difficulty in breathing, loss of sense of smell and taste, and diarrhea were put on the front burner. Advice was issued on the need to report to healthcare facilities if one experienced these symptoms in different combinations.

The public was also informed of the vulnerability of the elderly especially those with underlying conditions that can complicate the course of infection and the risk of transmission posed by the large pool of asymptomatic younger population. By November 2020, pregnant women, individuals with chronic kidney diseases, and children with sickle cell anemia have been added to the vulnerable list (CDC, 2020). A recent study linked obesity with increased hospitalization.

Countries were encouraged to do massive testing of their population in order to have accurate data to plan and to isolate the infected in a bid to limit the spread. As a duty to society, public health guidelines were put in place to slow the spread of the virus. The new normal of frequent hand washing with soap and water, use of alcohol-based sanitizers if water is not readily available, wearing of a face mask, avoiding crowded spaces, sneezing and coughing into the elbows, avoiding touching of the face with contaminated hands, and social distancing are now the passwords of social interactions and optimal living. Regular temperature checks are done before admission into venues for events with large population, schools, supermarkets, restaurants, etc. Self-isolation and quarantine of exposed individuals were recommended on contact with possible high-risk individuals or visits to high-risk areas. Aggressive lockdowns were issued as stopgap remedies to flatten the curve of infection. Massive awareness campaign was mounted by science advocates and major policy change enacted by different global governments to sensitize and enforce these guidelines for the well-being of society.

Research in the areas of possible and probable treatments and preventive vaccines went on overdrive to find the right molecules or combination of molecules that can be administered for relief of symptoms or keeping the infection out through improved immunity systems. Once progress was made in these directions, the information was released with regard to safe vaccines, the number of vaccine shots that can protect individuals, and possible side effects. Science also strenuously made the point that the vaccines need to be injected into the arms of the population and not kept unduly in storage facilities. They urged governments the world over to become more proactive in setting up vaccination centers and mount campaigns against vaccine hesitancy and conspiracy theories.

With the entrance of the vaccines and the possibility of bringing the pandemic around, scientists out of duty again had to break the not cheering news of the emergence of new variants of the virus with high transmissibility. These have been identified as three major variants, namely, B.1.1.7 with 23 mutations observed in patients in the UK, B.1.351 with multiple mutations in the S protein found in South Africa, and the P.1 with 17 mutations occurring in Brazil (Kamps & Hoffmann, 2021). Studies have shown that the UK variant may have caused more deaths than the original COVID-19 virus, while the available vaccines may be less effective against the Brazil variant. However, people are urged to take the vaccines and observe the containment guidelines, while studies are carried out to further unravel the nature and effects of these or any future variants.

The World Health Organization has been at the forefront of this duty to society working with a coalition of partners and scientists in the tracking of the pandemic, advising on critical interventions, and distributing vital medical supplies.

11.3   Beneficence

Another major principle is that researchers should not expose participants to unnecessary risks. At each point in time, the welfare of the research participant must be paramount, and the goal should strive for the benefits of the research to outweigh the risks. The current race for an effective vaccine is an example of science standing in the beneficence ethic by following all necessary protocols of vaccine development despite the urgency being experienced and the death toll rising from infections by the virus. Though there is enormous pressure from political circles and increasing numbers of infection, the vaccine journey is following the course of ensuring that safety of participants in trials and that end users are not sacrificed on the altar of efficacy and shortened processes. They also ensured that all the steps are being followed and data properly analyzed to show evidence-based trail and trend before approval can be given for mass usage. There are many strict protections in place to help ensure that COVID-19 vaccines are safe. Like all vaccines, COVID-19 vaccines are going through a rigorous, multistage testing process, including large (Phase III) trials that involve tens of thousands of people, especially volunteers from high-risk groups. Certain vulnerable groups are excluded such as lactating and pregnant women until the safety is ensured during these conditions (WHO, 2021a). Of course, this step supports the need to obtain informed consent from participants, and volunteers have enthusiastically enrolled to be part of the study.

There are about 48 candidate vaccines on the WHO draft vaccine landscape in November 2020, undergoing various stages of trials with many in Phase III clinical trials, and in early January 2021, 63 vaccine candidates were in clinical development against SARS-CoV-2 and 172 in preclinical development (WHO, 2021b). The announcements by Pfizer and Moderna of vaccine candidates with 90% and 95% effectiveness, respectively, in early December 2020 elicited excitement and optimism and hope of light at the end of the tunnel to protect a large number of the populace from the infection. However, caution was still exercised as the safety profiles of the vaccines were monitored and evaluated. The WHO insists that the approved protocol for vaccine development and authorization must be followed that once vaccines are demonstrated to be safe and efficacious, they must be authorized by national regulators, manufactured to exacting standards, and distributed (WHO, 2021a).

As of January 8, 2021, three vaccines had been approved in Europe and in the USA: BioNTech/Pfizer vaccine, Moderna vaccine, and University of Oxford/AstraZeneca. Three other vaccine candidates were also approved outside Europe and America. These are China (Sinopharm and the Beijing Institute of Biological Products vaccine), India (Covaxin by Bharat Biotech), and Russia (Sputnik-V Gamaleya Research Institute) (Kamps & Hoffmann, 2021). Common side effects on the administration of these vaccines include pain at the injection site, fatigue, headache, and fever, though some blood clot issues were raised in AstraZeneca/Oxford vaccine, leading to its suspension in some countries and advice not to be used in younger individuals. However, the European Union medical research team still urges European nations to use the AstraZeneca/Oxford vaccine since the benefits outweigh the risks.

As of February 18, 2021, at least seven different vaccines across three platforms have been rolled out in countries. At the same time, more than 200 additional vaccine candidates are in development, of which more than 60 are in clinical development (WHO, 2021b). Vulnerable populations in all countries are the highest priority for vaccination, and these include frontline workers, the elderly, and those with underlying conditions.

11.4   Integrity

The ethic of integrity was captured; thus, "Researchers should demonstrate honesty and truthfulness. They should not fabricate data, falsify results, or omit relevant data. They should report findings fully, minimize or eliminate bias in their methods, and disclose underlying assumptions" (Weinbaum et al., 2019).

This principle drove the late Dr. Li Wenliang, a Chinese ophthalmologist (Plate 11.1) from Wuhan, to alert his colleagues on a new disease similar to SARS he has

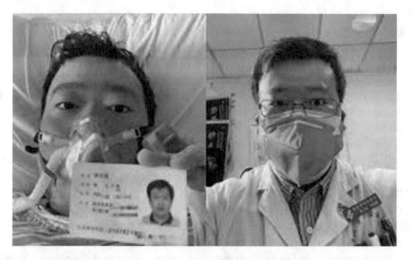

Plate 11.1 Photograph of Dr. Li Wenliang, a Chinese ophthalmologist

seen patients show up with at his hospital. He wanted this information to go out with an advice to friends and families to wear protective gears. He was harassed by the repressive officials of his country and later died from the infection. But the truth he told which was termed rumor was later vindicated.

In our own clime of Nigeria, Dr. Chikwe Ihekweazu, the Director General of the Nigeria Centre for Disease Control (Plate 11.2), has been a powerful voice of scientific truth constantly drawing attention to the need to observe guidelines and improve facilities at surveillance and response, he also evolved an efficient supply chain and appropriate use of personal protective equipment (PPE) especially to health workers. He showed the example of self-isolating for 2 weeks on his return from an

Plate 11.2 Photograph of Dr. Chikwe Ihekweazu, DG. NCDC

official assignment to a high-risk country, China. He is still advocating on the need to still follow the guidelines on prevention and containment, even though vaccines have arrived the country and may induce some of carelessness.

Dr. Tedros Adhanom Ghebreyesus, the Director General of the World Health Organization (WHO) (Plate 11.3), has not failed to speak the truth to power in this pandemic. He has stood his ground calling out on the need not to politicize the virus and to the fact that there are no "magic solutions" to the outbreak but just hard work from leaders and society at all levels. Dr. Ghebreyesus clearly stated on October 26, 2020, that "Science continues to tell us the truth about this virus. How to contain it, suppress it and stop it from returning, and how to save lives among those it reaches. Countries that have followed the science have suppressed the virus and minimized deaths. Where there has been political division at the national level, where there has been blatant disrespect for science and health professionals, confusion has spread, and cases and deaths have mounted" (VOA, 2020). With the rollout of the vaccines, he has explained why vaccines must be administered at large scale across different populations, encouraged against vaccine hesitancy, and urged developed countries not to start enacting laws on vaccine passports because the vaccines are still largely unavailable especially in developing countries.

A most respected and steady voice of science and the truth it stands for is Dr. Anthony Fauci (79 years), Director of the National Institute of Allergy and Infectious Diseases, USA, who has been waging an unending battle with the government and people of the USA to key into the simple guidelines that will help to keep the infection numbers low. His prediction in June 2020 that America will see unprecedented surge in infection levels and number of deaths from October if they continued on the road of opening up the economy and refusing to wear masks has come to pass and cases still rising.

He joins the scientific community to be cautiously optimistic of vaccines and their effects on this virus. He said, "The Calvary is coming but the Calvary is not yet here. Let the hope of a vaccine motivate us more to strictly follow the guidelines"

Plate 11.3 Dr. Tedros
Adhanom Ghebreyesus,
the Director General of the
World Health Organization
(WHO)

Plate 11.4 Photograph of
Dr. Anthony Fauci
(79 years), Director of the
National Institute of
Allergy and Infectious
Diseases, USA

(CNN, 2020). The Calvary has arrived and Dr. Fauci (Plate 11.4) joins in being one of the firsts to take the vaccine while advocating the establishment of more vaccination centers across the USA.

Many scientists have upheld this ethic during this pandemic, clearly and honestly stating the facts of this pandemic as they understood it. They did not hesitate to state that there were many things still evolving given the fact that they were dealing with a novel virus and were humble to say I do not know for now or for sure. Data from different laboratories and groups are being reported, processed, and analyzed to give a holistic view to the unfolding scenario. Reporting with integrity has been clearly demonstrated with announcements on vaccine development, vaccine effectiveness and efficacy, and vaccine administration and duration of protection the vaccines confer. Each step has been marked with transparency, honest reporting of findings, and humility to admit there are still gray areas.

Therefore, in this cautious optimistic state, the WHO joins other scientists to advise that "Safe and effective vaccines will be a game changer: but for the foreseeable future we must continue wearing masks, physically distancing and avoiding crowds. Being vaccinated does not mean that we can throw caution to the wind and put ourselves and others at risk, particularly because it is still not clear the degree to which the vaccines can protect not only against disease but also against infection and transmission" (WHO, 2021b). The emergence of the UK, Brazil, and South African variants supports this advice as studies are ongoing to understand if current approved vaccines are effective against these variants.

Recently, Italian cancer researchers unexpectedly discovered that some participants of a lung cancer screening trial showed SARS-CoV-2 antibodies dating back to September 2019 long before their index case of COVID-19 in January 2020 (GHN, 2020). This finding could reshape the history of the pandemic as the Chinese officials are already implying that this indicates that the virus did not originate in Wuhan but that Wuhan only sounded an early alarm to the presence of the disease.

11.5   Nondiscrimination Ethics

Attention has also been drawn to the gender issues arising from the pandemic. Many frontline health workers are females and may be at higher risk of being infected and taking the infection home to their families as they carry out their caregiving chores at home. Women also provide the majority of unpaid care work, including health-care work, in the home. The additional care burden associated with COVID-19 needs to be recognized and should be incorporated into policymaking and response measures. Therefore, effort must be made to ensure that PPEs which adequately protect the females are provided to these workers.

Data also show higher mortality rate among males than females which is another pressure point for the females as they lose spouses, partners, and male family members that provide necessary support in the household. The restrictions and attendant economic downturns have also escalated gender-based violence with women and children being at the receiving end, and science advocates are calling for policies that can protect and bring succor to these victims.

Member States of the WHO and their partners have been called upon to include responses to violence against women, and particularly intimate partner violence, as an essential service within the COVID-19 response (WHO, 2020). Also focused attention on the pandemic has hampered access to reproductive health interventions and other routine vaccinations and monitoring of intervention effort in some morbid conditions. These could lead to some disadvantages to the females such as maternal mortality, unwanted pregnancies, and resurge of preventable infections, hence the need for shifts in policies to address these needs.

The need for equitable access to vaccines has been highlighted by scientists especially given the attitude of rich nations to the acquisitions of vaccines. It has been observed that lack of equitable distribution of vaccines may undermine the great relief felt at the unequalled pace of production, approvals, and early rollout of the vaccines which are great public health achievements. There is an increase in vaccine rollout in high-income countries, while little or none is being done in low-to middle-income countries which points to a moral failure. The COVAX coalition for vaccine procurement and distribution was bought into by 190 countries and would enable vaccination of two billion of the most vulnerable groups in society. However, in this pandemic, high-income countries are doing direct purchase of vaccines bypassing the COVAX corridor and often in very large quantities, and they constitute just 16% of the world's population. Countries have put their survival first and no longer want to readily play their brother's keeper, but scientists warn seriously that apart from the moral question this attitude poses serious public health risks.

A strident voice in this area is the American Society of Tropical Medicine and Hygiene (ASTMH) that captured it; thus, in their March 2021 statement from the society, "A large number of unvaccinated people is a potential driver of mutations and dangerous variants, prolonging the pandemic, increasing global mortality and morbidity and impeding economic recovery and return to normal life" (ASTMH, 2021). Thus, no one is protected until we are all protected.

Scientists advocate the following positions, a summary of which is captured (ASTMH, 2021) with the following points on the way forward against nondiscrimination. COVAX countries are advised to abide by the strategic guidelines set forth in global interest. COVAX should be fully funded with open and transparent procuring schemes involving free or reduced cost vaccines to nations in need. There should be effort at introducing innovative delivery strategies for equitable access. Countries should invest on research with a view to produce vaccines with less stringent storage requirements preferably in single dose and that can protect against new variants. The open nature of science in information sharing should be encouraged among collaborators and regulators for faster access equitably to safe and effective new products once they pass quality control. Currently developing nations are calling for patent waiver to enable many more manufacturers to produce the vaccines in larger quantities to meet the demand worldwide. This request is before the World Trade Organization, but fears are high that this request will be blocked by the USA and the UK.

11.6 Privacy and Confidentiality

Issues on privacy and confidentiality violations have arisen given the digital modes employed in addressing the pandemic including contact tracing, cell phone monitors, and private data being shared online as people worked remotely. The unfettered access to private mobile technologies may pose a violation, but current arguments are that these data are needed and can be employed in the public interest as healthcare needs currently outweigh privacy concerns. One of the responses to the COVID-19 pandemic in a bid to slow the rate of transmission by public health practitioners and governments was to involve technology in contact tracing to estimate how far and wide an infected individual has come in contact with other people. It is also used in notifying such contacts that they may have been exposed to infection and need to take necessary precautions. This led to many contact tracing apps emerging into the public space. Two main approaches are used by these apps to log in data either the centralized or decentralized approach. In a centralized approach (used by contact tracing apps in the UK, Singapore, and Australia), a user's data is uploaded to a main server where public health authorities can review and analyze it. In a decentralized approach (used in Holland's contact tracing app), data remains on the user's mobile device with a "minimal amount of information uploaded to the server" (Lewis & Bockius, 2020).

Apps that use a centralized approach have more privacy risks (as data could be stolen or used for other purposes), but some say it gives authorities better insight into the spread of the virus. Apps that use a decentralized approach are more privacy friendly, as the data stays on users' devices. The capture of the location data of an infected individual seems like over-monitoring, and lack of anonymity could lead to stigmatization (Lewis & Bockius, 2020).

11.7 Conclusion

Looking through the lens of the pandemic, science seems to have tried to maintain ethical standards even under extreme pressure. Guidelines were put in place to slow the rate of transmission and to give health systems a window of operation without being overwhelmed. The only disturbing element in the operations was the intransigent human being who loved his pleasures more than he respected the voice of science and the mechanisms of nature. If only we all listened and obeyed especially the young who thought they were invincible (but data show that they form a significant percentage of the dead, about 20% in America), the outcome would have been better.

Hopefully, lessons have been learned on many sides from the government, the people, and scientists. There were missteps by scientists in the early days, some of which went against the tenets of ethics that saw even published works retracted and flip flops in the details of the mode of transmission but the tenacity to perform a duty to society at lower risks with integrity and honesty while protecting the confidentiality was not in doubt. It shone through. Science must not forget that it only uncovers what is already in existence and must look to nature to understand its laws and become a cooperating link, a voice of truth.

References

American Society for Tropical Medicine and Hygiene. (2021). *Statement from the society March 2021*. www.astmh.org. Accessed 3 Mar 2021.

Cable News Network Thoughts on vaccines December 5, 2020. www.cnn.com Accessed 3 Dec 2020.

Centers for Disease Control. (2020). *Vulnerable groups for COVID-19*. www.cdc.gov 2020. Accessed 20 Nov 2020.

Global Health Now (GHN) News October 2020: Italian Cancer Researchers uncover SARS-CoV-2 Antibodies. www.globalhealthnow.org. Accessed 7 Nov 2020.

Kamps, B. S., & Hoffmann, C. (2021). COVID reference www.CovidReference.com Edition 2021.6. Accessed 11 Mar 2021.

Lewis, M., & Bockius, L. L. P. (2020). *Data privacy issues in COVID-19 contact tracing apps*. www.lexology.com. Accessed 16 Mar 2021

Voice of America News. (2020). *WHO director warns "Don't politicize COVID-19 pandemic"*. www.voanews.org October 26, 2020 Accessed 18 Dec 2020.

Weinbaum, C., Landree, E., Blumenthal, T., & Gutierrez, C. (2019). *Ethics in scientific research* (p. 118). Publisher Rand Corporation.

World Health Organization. (2020). Gender and COVID-19 advocacy brief 14/5/2020. www.who.int. Accessed 20 Nov 2020.

World Health Organization. (2021a). https://www.who.int/news-room/q-a-detail/coronavirus-disease-(covid-19)-vaccines. Accessed 11 Mar 2021.

World Health Organization. (2021b). https://www.who.int/emergencies/diseases/novel-coronavirus-2019/covid-19-vaccines. Accessed 8 Mar 2021.

Chapter 12
Time and Resource Optimization for Career Advancement by Women Scientists from Resource-Poor Settings

Temitope Olawunmi Sogbanmu and Temitope Olabisi Onuminya

12.1 Introduction

Underdeveloped, developing countries, aboriginal, immigrant and refugee settlements in developed countries, among others, are often characterized as resource-poor settings (Dickens & Cook, 2003). Women scientists from such settings grapple with several challenges in maximizing their productivity alongside the gender-specific challenges as a female scientist. Time is a precious commodity which is universal in its span for everyone globally. An individual has 24 h each day to utilize as he or she wills. However, how this time is utilized by various persons and the resultant outcomes differ. As academics and scientists, the use of time for various activities majorly research and teaching depend on the individual scientist. Women scientists in particular face the dilemma of utilizing time productively to advance their research and invariably careers while managing the peculiarities of being a female or woman in science (Fathima et al., 2020). This is particularly so because the 'years of intensive research and long hours in the laboratory, usually required to establish an independent research career path, often coincide with the period in which most women tend to start families' (Hansen, 2020). Also, women are generally more likely to be the primary caregiver for young children as well as ill or

T. O. Sogbanmu (✉)
Ecotoxicology and Conservation Unit, Department of Zoology, Faculty of Science, University of Lagos, Lagos, Nigeria

TETFund Centre of Excellence on Biodiversity Conservation and Ecosystem Management (TCEBCEM), University of Lagos, Lagos, Nigeria
e-mail: tsogbanmu@unilag.edu.ng

T. O. Onuminya
TETFund Centre of Excellence on Biodiversity Conservation and Ecosystem Management (TCEBCEM), University of Lagos, Lagos, Nigeria

Department of Botany, Faculty of Science, University of Lagos, Lagos, Nigeria

© The Author(s), under exclusive license to Springer Nature Switzerland AG 2022
E. O. Nwaichi (ed.), *Science by Women*, Women in Engineering and Science, https://doi.org/10.1007/978-3-030-83032-8_12

ageing members of their families which force them to take career breaks or part-time appointments (Hansen, 2020). Other factors such as lack of support networks (including role models, mentors and sponsors), reduced access to resources and technology as well as negative perceptions have also been identified as important challenges, thereby leading to reduced productivity for women in science, technology, engineering and mathematics (STEM) (Vaughan & Murugesu, 2020).

Career advancement is measured by various metrics (Schimanski & Alperin, 2018). In academics and sciences, metrics such as research outputs (publications, patents, products), grants, fellowships, awards, teaching load, number of supervised students and administrative duties, among others, are utilized to determine tenure and promotion (Bonaccorsi & Secondi, 2017). In resource-poor settings (i.e. areas, countries, environments that lack or have little infrastructure, facilities or enabling environment to support maximum productivity), career advancement is often a challenge especially if adjudged at par with global benchmarks (Malelelo-Ndou et al., 2019). For women scientists and academics from these resource-poor settings, career advancement and ultimately global impact is often a mirage or an achievement by a few women (Mukhwana et al., 2020). However, in recent times, the United Nations Sustainable Development Goals (UN SDGs) aiming to 'leave no one behind' have proposed several corresponding targets and indicators which aim to address gender issues and mainstreaming for sustainability (UN Women, 2019; Herbert et al., 2020). The UN SDGs which are aimed to be realized by 2030 promises to put an end to barriers that prevent women and girls from realizing their full potential. While there has been some progress over the decades, significant challenges lie ahead especially in resource-poor settings such as presented in sub-Saharan Africa as subject segregation persists across regions, reflecting traditional expectations of gender roles (UN Women, 2020). In particular, women are persistently under-represented in science, technology, engineering and mathematics (STEM), which can have long-term effects not only on gender equality but also on economic development. Ultimately, tackling gender barriers and realizing gender equality in real terms involves addressing gender-based discriminatory norms, strengthening women's agency and exercising their rights in practice (Stuart & Woodroffe, 2016). A strong need in achieving these goals and targets is for capacity building on time and resource optimization particularly for women in sciences from resource-poor settings.

Optimization, a term often applied in technology or businesses, is concerned with the use of a resource, method or process in order to derive the maximum benefit possible within the ambits of the resource capacity (APM, 2019). Resource optimization, therefore, is an essential step in achieving career development in every field especially in science. It requires proper planning and management of time and other available resources in order to fulfil one's goals. Hence, with regard to career advancement, time and resource optimization may be referred to as the maximization of the period, available facilities and structures within one's environment to achieve the highest productivity possible. In this chapter, we posit some personal time- and experience-tested ways to optimize time and available resources as women in science from a resource-poor setting (public university in a developing

African country, Nigeria). Further, we reviewed literature culled from Google Scholar on the subject and key terms with a view to enrich the discourse with evidence-based approaches for career advancement of women scientists in resource-poor settings

12.2 Peculiarities of Women Scientists in Resource-Poor Settings

There are usually few women scientists at the top level/professorial/management positions in tertiary institutions, government agencies and industries in resource-poor settings (Downs et al., 2014). Where they are found, they are often older in age compared to their male colleagues or counterparts in 'resource-rich settings'. They are often found to successfully supervise few postgraduate (particularly PhD) students. Women scientists are often not principal investigators or research group leaders. Overt (obvious or open) discrimination, gender stereotypes of women such as housewife(ves), extended family responsibilities, 'culture' which often require women to be 'back benchers' or not overly ambitious, are some of the stereotypes that may be peculiar to women scientists in resource-poor settings (Ogunsemi et al., 2010). The inability to undertake research collaborations, fellowships, visits and sabbaticals 'abroad' or outside of their localities also poses challenge to women's career development especially from resource-poor settings (Prozesky & Beaudry, 2019). This may be due to poor research(er) visibility, lack of exposure, resource burden, lack of or poor mentorship, lack of access to or information about opportunities and lack of or low negotiation skills/influence over available (limited) resources at institutions in resource-poor settings, among others. In view of the foregoing, it is expedient for women in resource-poor settings to chart a course towards optimization of time and resources in their environment for effectiveness and personal development.

12.3 Approaches to Time Optimization for Career Advancement

Approaches for time optimization generally assume that resources are unlimited and that cost is not of primary concern. Although this is not applicable in most resource-poor settings, it helps in ensuring that deadlines are met and the deliverables of a project are not compromised. However, this may not be applicable in resource-poor settings with limited resources. Hence, to ensure career advancement for women scientists in resource-poor settings, it is important to ensure prioritization of task that one might engage in at different times. A useful tool for this is the

Fig. 12.1 Project
optimization (Uyttewaal,
2005)

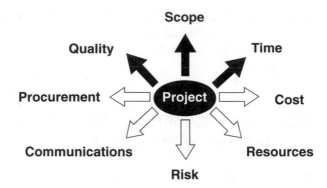

Microsoft Office Project 2003 which ensures dynamic scheduling of different aspects of one's project for a highly productive outcome (Fig. 12.1).

For example, as a woman scientist who is set to achieve a particular goal, one would need to do an analysis of the different components leading to the achievement of the goal to effectively optimize time. This can typically be achieved using the critical path approach in which one assesses the critical task needed to drive the time set for the achievement of the goal, and this can be done using Microsoft packages. Similarly, time optimization applications such as Microsoft Sticky Notes (Fig. 12.2) can be utilized to prioritize task and for proper scheduling. Further, setting annual/half-year/quarterly/weekly goals with a plan of activities (Table 12.1) is another way to optimize one's time (Brigette, 2004). Also, utilizing and optimizing the time and calendar applications on personal computers, desktop computers and mobile phones, among others, are quite useful for setting daily reminders.

12.4   Ideas on Resource Optimization

Resource optimization is particularly useful where the available resources for achieving a task are limited. Generally, resource optimization techniques include the following (Fig. 12.3): resource levelling (adjusting the start to finish dates based on the constraint of the resources without overburdening the individual) and resource smoothing (adjusting the activities of a scheduled project by redistributing resources while maintaining the timeline).

For women scientists aiming for career advancement in resource-poor settings, it is important to learn how to maximize the resources available in order to achieve career advancement within the time we have set for ourselves. This we can do without placing too much burden on ourselves or other aspects of our lives. Some of the useful tips in achieving resource optimization for career advancement in a resource-limited environment as a woman in science include, but are not limited to, the following.

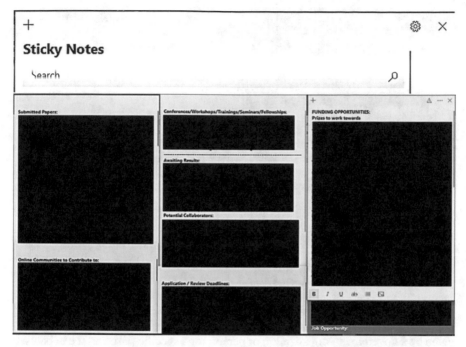

Fig. 12.2 Sample personalized sticky notes (Microsoft Sticky Notes; Sogbanmu, 2021)

Table 12.1 Sample template for annual/half-year/quarterly goals and targets

S/N	2021 goals/targets	Q1			Q2			Q3			Q4		
		Jan	Feb	Mar	Apr	May	June	July	Aug	Sept	Oct	Nov	Dec
1													
2													
3													
4													
5													

Money

This has been found to be a major limiting resource for women. So to optimize this, it is advised that you save a portion of your salary/income (e.g. 10–20%) for conference(s) attendance, capacity development/training courses, personal computer purchase and small field/laboratory equipment purchase, among others. Although this may seem insufficient, it will go a long way in preparing one for that desired advancement in the chosen career path. In addition, the need to vigorously source and compete for grants cannot be overemphasized as this is the bedrock for conducting meaningful research which can bring about innovation and development. In resource-poor settings, grantsmanship is a key option and may sometimes be the only available option. A *grant* is a quantity of money, i.e. financial assistance,

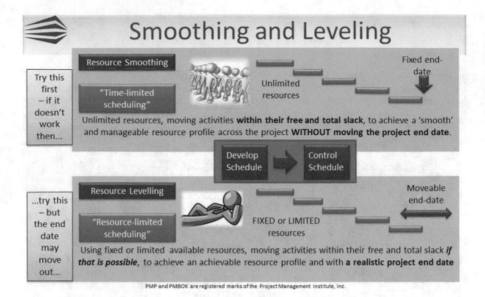

Fig. 12.3 Resource optimization techniques (Litten, 2020)

given by a government, organization or person for a specific purpose. It can be used to support research activities, equipment purchase, training, conference attendance and so on. The grant space is very competitive because the opportunities are limited. Understanding the chances of success is really important before applying for a grant, hence the need to use data to understand where to put one's resources. To succeed in grant applications, it is important to gain insight into the complicated funding landscape, discover hidden opportunities and make informed decisions. This includes searching for funding opportunities, selecting relevant opportunities, building teams, tracking, reporting and managing the grant. As a woman scientist, to increase competitiveness, the following are advised:

- Do not rush the process: it is important to plan carefully what to apply for, how and when to apply.
- Consult with a pro: learning from others who have succeeded in winning grants is useful so as not to 'reinvent the wheel' in error making.
- Think locally: think of providing solution to a local problem.
- Focus on the outcomes.
- Include a logic model: this is an outline of different components of the project, aim, methods, outputs and outcomes.
- Prove your effectiveness: the proposal needs to highlight the strength of the team, success stories and track record.
- Establish a measurement plan: plans for monitoring and evaluation to ensure deliverables of the grant should be clearly spelt out in the proposal or proper accountability.
- Ensure quality: the proposal must show value for money.

Support Network

This is another limiting resource especially as it relates to laboratory facilities, mentorship, goods, collaborations and visibility. Hence, optimization through identification and utilization of free online resources to increase research(er) visibility and
collaborative opportunities such as personal website/blog builder (e.g. WordPress –
see https://tosogbanmu.wordpress.com/) and research profile pages (Scopus,
Google Scholar, ResearchGate, Kudos, Publons, Academia, GitHub and so on) are
necessary. Also, there is a need to identify and work with collaborators with facilities both within and outside your university or country as well as actively utilize and
engage the networks available through these platforms especially ResearchGate.
Develop innovative research that can be conducted with available resources.

Engage and Get Involved

Success is never really achieved in silo; hence, there is a need to interact with and
get involved with other women scientists nationally, regionally and globally for
adequate support. This can be achieved by joining and getting involved in the activities of various organizations/networks of top-level scientists/researchers and utilize
it. For example, join and actively participate in activities of national (e.g. National
Young Academies (NYA)), regional (The African Academy of Sciences (The AAS),
Next Einstein Forum (NEF), African Science Leadership Programme (ASLP),
among others) and international (Organization for Women in Science for the
Developing World (OWSD), Global Young Academy (GYA), The World Academy
of Sciences (TWAS) and so on) scientific networks.

Vitae Researcher Development Framework (RDF)

This is a tool (Fig. 12.4) that can help to steer and support the career advancement
plans of researchers including women in science. A simple personalized RDF is
shown in Table 12.2 that may be adapted to develop personal career advancement
goals and associated activities.

Identify and Maximize Opportunities

Familiarize yourself with the conditions of service at your institution. Do not shy
away from responsibilities as this helps you build the requisite skills and experience
needed for your eventual career advancement. Do not miss a promotion/annual

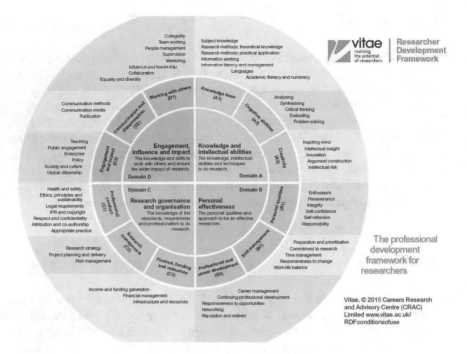

Fig. 12.4 Vitae Researcher Development Framework (https://www.wlv.ac.uk/research/research-policies-procedures%2D%2Dguidelines/vitae/vitae-rdf/)

increment. Prepare for and maximize the opportunities available therein such as maternity leave, research leave, annual leave (if you are unable to use it, bank it!), short-term and long-term leaves, training leave and sabbatical(s), among others.

12.5 Aiming for Global Impact and Relevance as Women in Science

Town and Gown Interaction – Translational Research

It is important particularly for researchers in resource-poor settings to engage various publics who their research has potential impact on for collaboration, funding, research uptake and ultimately impact.

Table 12.2 Sample personalized 2021 career goals based on Vitae Researcher Development Framework

Main domain	Sub-domains	2021 sub-domain goals	2021 objectives	Target	Timeline	Actions/ achievements
Domain A: Knowledge and intellectual abilities – the knowledge, intellectual abilities and techniques to do research	Domain 1: Knowledge base	Research methods: Practical application				
	Domain 2: Cognitive abilities	Critical thinking				
	Domain 3: Creativity	Innovation				
Domain B: Personal effectiveness – the personal qualities and approach to be an effective researcher	Domain 1: Personal qualities	Integrity				
	Domain 2: Self-management	Time management				
	Domain 3: Professional and career development	Continuing professional development				
Domain C: Research governance and organization – the knowledge of the standards, requirements and professionalism to do research	Domain 1: Professional conduct	Ethics, principles and sustainability				
	Domain 2: Research management	Research strategy				
	Domain 3: Finance, funding and resources	Income and funding generation				
Domain D: Engagement, influence and impact – the knowledge and skills to work with others and ensure the wider impact of research	Domain 1: Working with others	Supervision				
	Domain 2: Communication and dissemination	Publication				
	Domain 3: Engagement and impact	Public engagement				

Demand-Driven Research

Industry/policy priorities relevant to one's research within the locality, state or country should be addressed in research agenda of scientists. This will support uptake and relevance of the research outputs. One prime example is the United Nations Sustainable Development Goals (UN SDGs) with attendant targets and indicators. Women in science especially from resource-poor settings should identify which of these goals and targets align with their research expertise or vice versa and develop projects to feed into the indicators, hence global relevance. Similarly, the African Union Agenda 2063 Goals are another 'checklist' of research demands which women scientists can leverage on to align their research priorities and seek funds and/or collaborations. Sample personalized UN SDGs and AU Agenda 2063 Goals aligned with researcher priorities are shown in Table 12.3.

Capacity Building

Continuous and/or periodic participation in trainings, short courses, seminars, webinars some of which are free and accessible online is imperative to keep abreast with latest developments and techniques in one's research field.

Identifying Mentors and Being a Mentor

Seeking a mentor(s) either female or male who is experienced not necessarily in age but who has the skills and achievements that one desires in the medium to long term is important. Similarly, being a mentor to others such as peers, younger colleagues and even older colleagues is key. Endeavour to be open, willing to learn and willing to be of help as much as within your ability, time and expertise. Mentorship can also be indirect or from afar.

Project Yourself to the World (Research(er) Visibility)

This has been detailed in the previous section. It is highly important as women scientists particularly from resource-poor settings who are aiming for global impact. Expand your network and collaborations. Be bold and aim for top positions both locally, regionally and internationally.

Table 12.3 Sample Personalized Research Priority Plan based on Relevant Sustainable Development Goals and African Union Agenda 2063 Goals

Agenda 2063 Goals		SDG Goals		SDG targets		SDG indicators		Proposed projects
		Goal 3	Ensure healthy lives and promote well-being for all at all ages – good health and well-being	3.9	By 2030, substantially reduce the number of deaths and illnesses from hazardous chemicals and air, water and soil pollution and contamination	3.9.2	Mortality rate attributed to unsafe water, unsafe sanitation and lack of hygiene (exposure to unsafe water, sanitation and hygiene for all (WASH) services)	
Goal 7	Sustainability and climate resilience	Goal 6	Ensure availability and sustainable management of water and sanitation for all – clean water and sanitation	6.3	By 2030, improve water quality by reducing pollution, eliminating dumping and minimizing release of hazardous chemicals and materials, halving the proportion of untreated wastewater and substantially increasing recycling and safe reuse globally	6.3.1	Proportion of domestic and industrial wastewater flows safely treated	
						6.3.2	Proportion of bodies of water with good ambient water quality	
		Goal 11	Make cities and human settlements inclusive, safe, resilient and sustainable – sustainable cities and communities	11.6	By 2030, reduce the adverse per capita environmental impact of cities, including by paying special attention to air quality and municipal and other waste management	11.6.1	Proportion of urban solid waste regularly collected and with adequate final discharge out of total urban solid waste generated, by cities	

(continued)

Table 12.3 (continued)

Agenda 2063 Goals		SDG Goals		SDG targets		SDG indicators		Propose project
Goal 7	Sustainability and climate resilience	Goal 14	Conserve and sustainably use the oceans, seas and marine resources for sustainable development – life below water	14.1	By 2025, prevent and significantly reduce marine pollution, in particular from land-based activities	14.1.1	Index of coastal eutrophication and floating plastic debris density	
				14.2	By 2020, sustainably manage and protect marine coastal ecosystems and take action for their restoration in order to achieve healthy and productive oceans	14.2.1	Proportion of national exclusive economic zones managed using ecosystem-based approaches	

12.6   Case Study

Personally, I (T. Onuminya) am a woman scientist in the field of biodiversity conservation and ecosystem management. I began my career as a graduate fellow in the Department of Botany, University of Lagos in 2008. Having graduated with a First-Class BSc Degree in Botany, I was able to register for MPhil programme and successfully converted to a PhD programme in 2009. While I was not married at the time, a major challenge I encountered was finance. This was largely because I came into the programme without any work experience and not much money had been saved. However, I was able to succeed with my programme because I benefitted from the support received from my supervisor in the following areas:

First, he exposed me to several trainings and workshops which helped me to develop my capacity in grant writing. I also received a lot of guidance and encouragement on creating and having a proposal bank, which was constantly reviewed by other colleagues and senior colleagues during our research meetings which held every fortnight. With this, I was able to apply for and win a number of grants which helped in supporting my research work.

Second, I benefitted through conference attendance and exposure to a network of researchers in my field of expertise. This helped me greatly in building relationships with scientists in other institutions nationally and internationally and provided opportunities to access facilities which are otherwise not available to me in my institution.

Again, I was introduced to academic writing, publishing as well as ownership a publishing journal. With this, I was able to improve my academic writing skills and identify what to look out for in a journal before choosing for publication and so on. This has contributed immensely to my career advancement in the university as academic writing and publishing forms an integral part of the requirements. I must say that having good training, mentorship and exposure has greatly helped me to build and advance my career as a woman scientist in a resource-poor setting. Despite the limitations, I was able to successfully complete my PhD programme in flying colours, winning the National Doctoral Thesis award and graduating as the youngest PhD graduate ever produced in the University of Lagos in 2011.

Finally, getting married and having children some years down the line, I can say that it has been more demanding to pursue career advancement as a woman scientist. However, having good support from my spouse has been a good relief and support system for achieving advancement. It has in no small way assisted me in growing through the ranks, and my ability to maintain a disciplined lifestyle of constantly seeking for and applying for grants, keeping a network of researchers and building partnerships has helped in the process.

I therefore would like to encourage other women scientists in this clime to pursue grantsmanship, seek mentorship and build collaborations across the world to make their work a fruitful and effective one.

12.7 Conclusion

In this chapter, we have advanced strategies and time-tested approaches for optimizing time and resources as and for women scientists from resource-poor settings. Several of these approaches are also relevant for other women scientists who may not be from 'resource-poor settings'. We can be all we want to be and achieve if we believe in ourselves, have faith and work wisely. It is up to you to optimize your time and resources for impact as a woman in science.

Acknowledgements TOS acknowledges the Organization for Women in Science in the Developing World (OWSD), University of Port-Harcourt Chapter for the invitation to deliver a talk on 'Time and Resources Optimization for Women Scientists from Resource Poor Settings' in February 2021. This book chapter is borne out of the presentation.

References

APM (Association for Project Management) 7th edition (2019). *APM body of knowledge* (R. Murray-Webster, Ed.). APM. ISBN: 978-1-903494-82-0.

Bonaccorsi, A., & Secondi, L. (2017). The determinants of research performance in European universities: A large-scale multilevel analysis. *Scientometrics, 112*, 1147–1178.

Brigitte, J. C. C.-E. (2004). *Perceived control of time: Time management and personal effectiveness at work/ by: Technische Universiteit Eindhoven, –proefschrift.* ISBN 90-386-2147-7.

Dickens, B. M., & Cook, R. J. (2003). Challenges of ethical research in resource-poor settings. *International Journal of Gynaecology and Obstetrics, 80*(1), 79–86.

Downs, J. A., Reif, L. K., Hokororo, A., & Fitzgerald, D. W. (2014). Increasing women in leadership in global health. *Academic Medicine : Journal of the Association of American Medical Colleges, 89*(8), 1103–1107.

Fathima, F. N., Awor, P., Yen, Y.-C., Gnanaselvam, N. A., & Zakham, F. (2020). Challenges and coping strategies faced by female scientists – A multicentric cross sectional study. *PLoS One, 15*(9), e0238635.

Hansen, D. S. (2020). Identifying barriers to career progression for women in science: Is COVID-19 creating new challenges? *Trends in Parasitology, 36*(10), 799–802.

Herbert, R., Falk-Krzesinski, H., & Plume, A. (2020). *Sustainability through a gender lens: the extent to which research on UN Sustainable Development Goals (SDGs) includes sex and gender consideration.* https://doi.org/10.2139/ssrn.3689205.

Litten, D. (2020). *Resource optimization techniques.* Retrieved from: https://www.projex.com/resource-optimization-techniques/ Accessed 31 Mar 2021.

Malelelo-Ndou, H., Ramathuba, D. U., & Netshisaulu, K. G. (2019). Challenges experienced by health care professionals working in resource-poor intensive care settings in the Limpopo province of South Africa. *Curationis, 42*(1), 1–8.

Mukhwana, A. M., Abuya, T., Matanda, D., Omumbo, J., & Mabuka, J. (2020). *Factors which contribute to or inhibit women in Science, Technology, Engineering, and Mathematics in Africa.* Published by the African Academy of Sciences, Nairobi, Kenya.

Ogunsemi, O. O., Alebiosu, O. C., & Shorunmu, O. T. (2010). A survey of perceived stress, intimidation, harassment and well-being of resident doctors in a Nigerian Teaching Hospital. *Nigerian Journal of Clinical Practice, 13*(2), 183–186.

Prozesky, H., & Beaudry, C. (2019). Mobility, gender and career development in higher education: results of a multi-country survey of African Academic Scientists. *Social Sciences, 8*(6), 188.

Schimanski, L. A., & Alperin, J. P. (2018). The evaluation of scholarship in academic promotion and tenure processes: Past, present, and future. *F1000 Research, 7*, 1605.

Stuart, E., & Woodroffe, J. (2016). Leaving no-one behind: Can the Sustainable Development Goals succeed where the Millennium Development Goals lacked? *Gender and Development, 24*(1), 69–81.

UN Women. (2019). *Progress on the Sustainable Development Goals: The gender snapshot 2019.* https://dspace.ceid.org.tr/xmlui/bitstream/handle/1/1128/progress-on-the-sdgs-the-gender-snapshot-2019-single-pages-en.pdf?sequence=1&isAllowed=y Accessed 1 Apr 2021.

UN Women. (2020). *sdg-report-fact-sheet-sub-saharan-africa.* UN Women.

Uyttewaal, E. (2005). *Dynamic scheduling with Microsoft Office Project 2003: The book by and for Professionals* (755 pp). J. Ross Publishing, International Institute for Learning, Inc.

Vaughan, A., & Murugesu, J. A. (2020). Science's institutional racism. *New Scientist, 3288*, 14–15.

Chapter 13
The Urgent Need to Scale Enlightened Feminine Leadership for Conscious Evolution

Eliane Ubalijoro

13.1 Introduction

Data from many sources are pointing to the numerous effects of COVID-19 on women researchers and academics work including how gender disparities have been amplified. Women take on more household work and more caretaking tasks despite also having to give their best in their profession at the same level as men in the same positions. Women represent less than 30% of the global science, technology, engineering, and mathematics (STEM) research workforce according to UNESCO (2019). These inequities are amplified by the reality that "we continue to observe twin deficits in women's representation as leaders in science – notably, in natural sciences, and engineering and maths – and in the integration of gender analysis in these fields" (IDRC, 2021). Gender equity in STEM requires addressing the leadership gap while also ensuring that gender analysis helps us understand how to bring more women into STEM and keep them there along their careers. As both a female scientist and leadership scholar, I use my own journey as a scientist and leader to understand what barriers I have faced that other women may also face. I will also explore how cultivating feminine archetypes and their representation through African goddesses has helped me gained soft skills that have been critical to my navigating turbulence in my career journey. I will also discuss how these archetypes have guided me beyond the limited possibilities I had earlier in my career to see

E. Ubalijoro (✉)
Future Earth Canada Hub, Montreal, QC, Canada

Department of Geography, Planning and Environment, University of Concordia, Montréal, QC, Canada

Institute for the Study of International Development, McGill University, Montréal, QC, Canada
e-mail: eliane.ubalijoro@mcgill.ca

© The Author(s), under exclusive license to Springer Nature Switzerland AG 2022
E. O. Nwaichi (ed.), *Science by Women*, Women in Engineering and Science, https://doi.org/10.1007/978-3-030-83032-8_13

169

women who look like me in positions of leadership to dream beyond what my eyes could see.

13.2   Confidence Versus Competence

According to Katty Kay and Claire Shipman (2018), in their groundbreaking book, *The Confidence Code*, "professional success demands political savvy, a certain amount of scheming and jockeying, a flair for self-promotion and not letting a no stop you. Women often aren't very comfortable with that. Perhaps, deep down, we don't really approve of these tactics. Whatever the reason, we haven't been very good at mastering these skills, and that holds us back." Kay and Shipman studied leading innovative research across many fields from genetics behavior, gender to cognition. They also used their own experiences and those of women leaders to come to the conclusion that confidence (not competence) will be critical in tipping women's career forward or not. At the Vancouver Peace Summit 2009, the Dalai Lama stated that "the world will be saved by the western woman" (Lowen, 2019). I remember reading this and feeling very sad as an African woman from the Global South that this excluded me somehow. What I or the Dalai Lama did not know back then was that by 2035, half of the global workforce will be African. What I also do know now is that Africa's population is slated to reach 2.6 billion by 2050. What this means is that African women can deeply shape the future of humanity not only through how many children they birth but also through how they are able to nurture these children into adulthood to contribute to humanity. This will depend on how they are able to own their own power and live fulfilling thriving lives.

13.3   The Great Reset

The pandemic is the great reset we did not know we were waiting for. As we collectively face this crisis, there is also a collective opportunity to pivot and build forward stronger in ways that respect nature and nurture the amazing array of biodiversity on the planet. As a leadership educator and facilitator, I constantly ask myself what can this moment I want to run away from teach me. How can all the places that hurt be breathed in so I can breathe out something that brings healing to me and others?

Centuries of blind ignorance of natural, social and human capital that did not increase financial gains have brought humanity to its knees. The entire global economy has been affected by lockdowns, loss of life, and overwhelm in many health systems as the number of cases has grown. On April 1, 2021, the COVID-19 cases around the world had reached 129,454,440 with 2,827,426 deaths (Worldometers, 2021). This is the most catastrophic pandemic that has affected humanity in a century. The daily new cases are still rising, sending a clear message; we are not near to the end of this pandemic as we urgently race to vaccinate all of humanity. The data is showing that caring economies that put in place measures to halt the virus spread early are also the ones showing the fastest recovery. According to Bastian (2020), "handling this pandemic well is going to be about far more than containing the virus for several months. Success will also mean mitigating the socioeconomic consequences of the pandemic including racial and social disparities, achieving high enough vaccination levels if and when that time comes, meeting the potential tsunami of long-term health needs for people post-Covid and those in long-term care, and establishing excellent pandemic preparedness for the next one."

While COVID continues to hold us in fear, biodiversity loss keeps accelerating; the pandemic of racism is one humanity is struggling with as well. As an African woman who has studied science in North America from my undergraduate days to obtaining my PhD, I have lived the duality of my identity by birth and my career choices in the incredulity of some when I show up as a keynote speaker or in my own imposter syndrome. What I know is that the reset we are in today, calls for all, especially women to own their power, for all African women, especially scientists, to contribute to environmental, economic, and social transformational changes to end inequity and help make livelihoods sustainable everywhere on the planet. "Moments of existential crisis bear within them the ability to dream and imagine new possibilities" according to Mamphela Ramphele (2020). I would like to take this opportunity to dream with you reader about what empowerment would look like for all women, especially African women scientists.

"A recent Stanford Business School study shows that women who can combine male and female qualities do better than everyone else, even the men. How do they define the male qualities? Aggression, assertiveness, and confidence. The feminine qualities? Collaboration, process orientation, persuasion, humility" (Kay & Shipman, 2018). For millennia, these qualities have been embodied in mythology from all over the world through symbolic characters that convey different strengths we get explore through storytelling and belief systems often embedded in gods and goddesses we aspire to emulate. Archetypes are symbolic embodiments of character. According to Deepak Chopra (2020), there are seven feminine archetypes the world needs to cultivate urgently for conscious evolution to happen. My intent in sharing these archetypes is to promote embodied courage in women scientists like me. How do we cultivate the strengths we need to succeed professionally while also encouraging self-care? Without both, many of us become part of the worrying "leaky pipeline" of women who leave science education and careers in the prime of their lives. "No one in a position of power along the pipeline has consciously decided to filter women out of the STEM stream, but the cumulative effect of many separate but related factors results in the sex imbalance in STEM that is observed today" (Clark Blickenstaff, 2005).

13.4   Our Ancestors Are Forever Part of Us

The leaky pipeline needs to stop so the 9 years left to achieve the 17 Sustainable Development Goals can be achieved with all of humanity's help. This requires the left hand and the right hand working together. Having one hand tied behind one's back cannot do. That is what gender inequity continues to force humanity to do.

> We are endowed with a divine spark that never dies because it is connected to the source of all life. The dead are never dead. Our ancestors are forever part of us… We are inextricably related to our ancestors, who continue to live in present generations as guiding spirits. We stand as bridges to future generations (Ramphele, 2020).

Embracing archetypes that embody the greatness of our ancestors allows us to take our destiny in our hands and not let fate decide our future. Fate in the case of women scientists is all accumulated data that demonstrate how the odds are against us staying in our chosen STEM fields and growing there as leaders. Fate in the case of COVID-19 is the reality that zoonosis is on the rise and is responsible for 75% of infectious disease transmission between wildlife and humans. Zoonosis has been amplified because of the loss of over 50% of the biodiversity worldwide over the last 60 years. According to the Intergovernmental Science-Policy Platform on Biodiversity and Ecosystem Services (IPBES), "the health of ecosystems on which we and all other species depend is deteriorating more rapidly than ever. We are eroding the very foundations of our economies, livelihoods, food security, health and quality of life worldwide" (UN, 2019).

How do we consciously design a world where women thrive in STEM fields so their work can scale the services nature gives us freely and abundantly? We urgently need a world where all the abundance nature gives us is clearly accounted for and not ignored in the ways that have led to accelerating biodiversity loss, increased outbreaks of zoonosis, and brought humanity to its knees with the current pandemic. Reconnecting to the earth is needed. According to Rattan Lal, the 2020 Food Prize Winner, over 70% of African soils are degraded. What does that say about our sacred connection to the earth? Regenerating soils to health will increase the diversity of microorganisms; we can discover medicine from and make our ecosystems so resilient that they mitigate against climate change and contribute to food security. For centuries, African soils have seen the depletion of our resources, the tears of mothers suffering losses from slavery, colonization, and conflict. A reconnection to the sacred is needed while we also harness the power of science and technology to accelerate transformations that will allow humanity to live within planetary boundaries. Here are the related seven archetypes to cultivate:

- Motherhood
- Self-esteem
- Nature
- Wisdom
- Beauty
- Healing
- Care

I will go through each of these archetypes to explore how as women scientists today we can embody these archetypes. By embracing and living these symbolically, we can infuse the feminine in all our actions, relationships, and decision-making. We can be models for all people across gendered identity for the conscious leadership urgently needed to shift humanity on a course of fateful disaster to conscious design towards climate action, reverence for nature, and a better world for all.

1. Motherhood is associated with the Greek goddess Demeter. I encourage each of you to identify a deity/shero of yours (living, dead, or fictional) who embodies intuition, affection, nurturing, and tenderness. *The Akan people of West Africa regard Asase Yaa as Mother Earth, the earth goddess of fertility and the upholder of truth.*[1]

2. The Greek goddess Hera represents power/self-esteem. The most powerful Yoruba goddess Yemayah is the orisha of the oceans.[2] All flows to and from her. My African s-hero of power and self-esteem is Wangari Maathai. I had the fortune of meeting her in Montreal. Her quality of presence, her capacity to have navigated a career in Academia full of bias against women and still stand tall without any feeling of inferiority or superiority was combined with charisma, friendliness, curtesy, graciousness, respect, and good manners.

3. The goddesses Artemis and Diana represent nature in Greco-Roman mythology. *iNyanga, Zulu, moon goddess, Nomhoyi, Zulu, goddess of rivers and Nomkhubulwane, Zulu, goddess mother of fertility, rain, agriculture, rainbow and beer*[3] are all African deities of nature. Aja is the orisha patron of forest, keeper of the secrets of healing plants and of all animals who dwell in the forest. Female nature divinities steward nature conservation[5]. They connect with all life and are conscious of our extended body that we call the environment. Mawu, the creator/moon goddess from the Dahomey, is the only one able to give the breath of life or take it[5]. A nature goddess holds reverence for air. Clean air fills our lungs with each breath and allows photosynthesis in plants. She holds reverence for clean water, the blood that flows through our circulation and hydrates all life. The earth allows the catabolism that is critical to cleansing and incubation of life as our collective recycling body where waste can be turned into rich compost. This rich compost increases soil organic matter so critical to the insect and microbial life that supports the resilience of our plant crops and biodiversity through adversity. From Mawu, we can embrace the notion of circular economy and design a world where all waste become valuable resources like fertilizer, energy, and reusable entities. *Gleti, the moon goddess of Benin, is the mother of all stars*[5]. All our atoms are gifts from start dust. This atomic dust is present in

---

[1] Frances Romero (2011). ARE YOU MY MOTHER? Asase Yaa
http://content.time.com/time/specials/packages/article/0,28804,2066721_2066724_2066705,00.html

[2] Chinaroad Löwche (2008). African goddesses: http://www.lowchensaustralia.com/names/african-goddesses.htm

[3] Wikipedia (2020). Nature deities: https://en.wikipedia.org/wiki/List_of_nature_deities

the minerals on earth and in our bodies. These feminine divinities remind us to take care *of the personal body and the extended body we call nature.*[4] As we navigate the COVID-19 pandemic, we are all affected differently. Ssome of us are essential workers, asked to go out in the world every day to be of service to others. Some are working from home, and the boundaries between activity and rest can blur us into exhaustion. For others, the anxiety of work interruption, loss or academic delays can bring on anxiety. In all cases, we are asked to care for ourselves, our extended bodies with consciousness.

4. The goddesses Athena, Saraswati, and Isis are holders of wisdom. All art, science, literature, culture, and music are gifts from wisdom. We are at a time in the world when wisdom is urgently needed for political leadership. This is critical in times of crisis where common sense is often forgotten.

5. The goddess Aphrodite and the orisha Oshun both represent sensuality, aesthetics, love, sexuality, and beauty, Aphrodite. These deities, for all genders, hold the power of attraction. They allow us to see our own beauty through our brokenness, our fears, our anger, our anxieties, our vulnerabilities and overwhelm. They remind us to love ourselves so our love for others is as pure as possible. They remind us of the need to live without judgment and to cultivate forgiveness. They ask us to connect to our sadness, to dance our way back to joy and use all art forms to uplift civilizations. According to Research Professor of Social Work Brené Brown, "vulnerability is the birthplace of innovation, creativity, and change." How can Oshun energy encourage us to contribute to positive transformation for mother nature and humanity?

6. The Yoruba goddess Oya and Greek goddess Persephone both represent transformation. They both have the capacity to journey to the underworld and come back. When we invoke their power for transformation, we connect to our capacity to heal. We connect with our alchemical power to move energy in a positive way, where there was darkness. We walk into darkness and can be bringers of light where there is deep unconsciousness. This is very critical to social justice movements. Oya and Persephone allow us to work with the energy of light and shadow, to face evil and hold mercy. They give us strength to navigate uncertainty. In the alchemical sense, how do we transform a STEM world full of barriers for women into one full of opportunities to contribute "the gold" of our intellects, hearts, and souls?

7. The last feminine power we will explore is the power to take care of the home and the world as our family. In Sanskrit, "the world is my family" translates as *Vasudhaiva Kutumbakam.* The Greek goddess Hestia and the Igala goddess Nekpe represent the homemaker, the protector of family. Louise Mushikiwabo, former Rwandan minister of Foreign Affairs, now current Secretary General of Organisation internationale de la Francophonie, embodies this quality of care for her constituents. Whether she was reaching out to young couples in the Rwandan

---

[4] Deepak Chopra (2020). Invoking the Divine Feminine for Conscious Evolution and Leadership: https://www.youtube.com/watch?v=JayRXUvbfa8&list=WL&index=61&t=0s

diaspora to encourage them to care for their homes and their homeland or now championing the livelihoods of the 800 million French speakers in the world, her focus has always been on care. Her 2006 autobiography is entitled *Rwanda Means the Universe*, a play on the fact that in Kinyarwanda, World, Universe, and Rwanda are all the same. She embodies the Sanskrit *Vasudhaiva Kutumbakam*.

In a world where there are so many barriers to our success, cultivating hope as African women scientists can become a job on its own. My hope is that cultivating time to connect with spirit, with the archetypal strengths of our ancestors embodied through the goddesses described here can help our dreams and aspirations so great that the barriers set before us can become as close to insignificant as possible.

13.5   Conclusion

What the pandemic has shown us is how interconnected we all are. We live in One World, why not live a One World Consciousness? Creativity is needed in all work to capture this oneness and bring it to all we do in ways that serve mother earth, our power/self-esteem, nature, wisdom, love, transformation, and care. Finding ways to bring these archetypes to life in our day-to-day actions is critical to our collective contribution to elevating human consciousness. Deepak Chopra (2020) asks us to "imagine a government with a president/prime minister who represented in some shape of form at least all these seven archetypes." We urgently need the scaling of conscious leadership that embraces these archetypes. We need to be the change we want for the world so let us practice being good listeners, growing our emotional intelligence, living with full awareness, being action oriented from our spaces of power, "taking responsibility, having integrity, having a heart calling, and changing the world. Imagine that having these seven archetypes as the cabinet of a great leader and extending that in the ecosystem in a distributed leadership" (Chopra, 2020). Will you join me on this journey?

References

Bastian, H. (2020). *What the data really says about women leaders and the pandemic*. https://www.wired.com/story/what-the-data-really-says-about-women-leaders-and-the-pandemic/

Blickenstaff, J. C. (2005). Women and science careers: Leaky pipeline or gender filter? *Gender and Education, 17*, 40369–40318. https://doi.org/10.1080/09540250500145072

Chopra, D. (2020). *Invoking the divine feminine for conscious evolution and leadership*. https://www.youtube.com/watch?v=JayRXUvbfa8&list=WL&index=61&t=0s

IDRC. (2021). *The Gender in STEM Research Initiative: Advancing gender analysis and women's leadership in STEM fields*. Call for Expressions of Interest. https://www.idrc.ca/en/gender-stem-research-initiative-advancing-gender-analysis-and-womens-leadership-stem-fields-call

Kay, K., & Shipman, C. (2018). *The confidence code: The science and art of self-assurance – What women should know*. Harper Business, an imprint of Harper Collins Publisher.

Lowen, L. (2019). *The Dalai Lama – "The world will be saved by the western woman"*. https:// www.thoughtco.com/dalai-lama-world-saved-western-woman-3971297

Ramphele, M. (2020). Chapter 5: Ubuntu: The dream of a planetary community. In P. Clayton, K. M. Archie, J. Sachs, and E. Steiner (Eds.), *The new possible: Visions of our world beyond crisis* (p. 45). Cascade Books, an Imprint of Wipf and Stock Publishers. Kindle Edition.

UN. (2019). *UN report: Nature's dangerous decline 'unprecedented'; Species extinction rates 'accelerating'*. https://www.un.org/sustainabledevelopment/blog/2019/05/ nature-decline-unprecedented-report/

UNESCO. (2019). *Women in science*. http://uis.unesco.org/en/topic/women-science

Worldometers. (2021). *COVID-19 coronavirus pandemic*. https://www.worldometers.info/corona-virus/. From last update as of April 01, 2021, 02:47 GMT

Correction to: Research Collaborations for Enhanced Performance and Visibility of Women Scientists

Chioma Blaise Chikere, Memory Tekere, Beatrice Olutoyin Opeolu, Gertie Arts, Linda Aurelia Ofori, and Ngozi Nma Odu

Correction to:
Chapter 4 in: E. O. Nwaichi (ed.), *Science by Women*, Women in Engineering and Science, https://doi.org/10.1007/978-3-030-83032-8_4

This chapter was inadvertently published without updating the following error:

The affiliation of the author "Chioma Blaise Chikere" was erroneously published in chapter 4. This has now been updated in this corrected version.

The updated version of this chapter can be found at
https://doi.org/10.1007/978-3-030-83032-8_4

E. O. Nwaichi (ed.), *Science by Women*, Women in Engineering and Science,
https://doi.org/10.1007/978-3-030-83032-8_14

Index

Printed in the United States
by Baker & Taylor Publisher Services